D0849115

Hydrocracking and Hydrotreating

John W. Ward, EDITOR
Union Oil Co. of California

Shai A. Qader, EDITOR
Burns and Roe Industrial Service Corp.

A symposium sponsored by
the Division of Petroleum
Chemistry, Inc. at the
169th Meeting of the
American Chemical Society,
Philadelphia, Penn.,
April 9, 1975

ACS SYMPOSIUM SERIES **20**

AMERICAN CHEMICAL SOCIETY

WASHINGTON, D. C.   1975

Library of Congress CIP Data

Hydrocracking and hydrotreating.
  (ACS symposium series; 20 ISSN 0097-6156)

  Includes bibliographical references and index.

  1. Cracking process—Congresses. 2. Petroleum—Refining—Congresses. 3. Hydrogenation—Congresses.

  I. Ward, John William, 1937-   II. Qader, S. A.
III. American Chemical Society. Division of Petroleum
Chemistry. IV. Series: American Chemical Society. ACS
symposium series; 20.
TP 690.4.H9   665'.533   75-33727
ISBN 0-8412-0303-2   ACSMC8 20 1–168

ACS Symposium Series

Robert F. Gould, *Series Editor*

FOREWORD

The ACS Symposium Series was founded in 1974 to provide
a medium for publishing symposia quickly in book form. The
format of the Series parallels that of the continuing Advances
in Chemistry Series except that in order to save time the
papers are not typeset but are reproduced as they are submitted by the authors in camera-ready form. As a further
means of saving time, the papers are not edited or reviewed
except by the symposium chairman, who becomes editor of
the book. Papers published in the ACS Symposium Series
are original contributions not published elsewhere in whole or
major part and include reports of research as well as reviews
since symposia may embrace both types of presentation.

CONTENTS

PREFACE

The efficient upgrading of energy resources to desirable products is an important national goal. This volume contains a series of papers concerned with the hydroprocessing of petroleum stocks, shale oil, and coal. Subjects covered range from catalyst structure to product properties. The hydroprocessing routes discussed offer excellent means of selectively producing the desired fuels and the means of reducing potential polluting contaminants from fuels and their precursors.

Union Oil Co. of California
Brea, Calif.

JOHN W. WARD

Burns and Roe Industrial Services Corp.
Paramus, N.J.
September 4, 1975

SHAI A. QADER

The Influence of Chain Length in Hydrocracking and Hydroisomerization of *n*-Alkanes

JENS WEITKAMP

Engler-Bunte-Institute, Division of Gas, Oil, and Coal, University of Karlsruhe, D-75 Karlsruhe, West Germany

Since 1960 catalytic hydrocracking has received considerable importance in petroleum refining where it is used mainly for the production of gasoline, jet fuel, middle distillates and lubricants. Its outstanding advantages are flexibility as well as high quality of the products. In the course of its commercial growth the chemistry of hydrocracking over various types of bifunctional catalysts has been scrutinized with model hydrocarbons, most commonly with alkanes. Several reviews on this subject are now available (1-3) showing that the product distributions are markedly influenced by the relative strength of hydrogenation activity versus acidity of the catalyst.

A unique type of hydrocracking associated with an utmost degree of product flexibility may be attained with bifunctional catalysts of both high acidity and especially high hydrogenation activity which have been counterbalanced carefully. The term "ideal" hydrocracking has been introduced (2) to characterize the reactions of n-alkanes with hydrogen on such catalysts. Typical features of ideal hydrocracking of long chain alkanes include:

1. low reaction temperatures
2. the possibility of high selectivities for isomerization
3. the possibility of pure primary cracking

all being in contrast to catalytic cracking over monofunctional catalysts. Actually, both reactions may be looked upon as pure mechanisms of cracking with many intermediate mechanisms between them. Such a concept has been shown to be fruitful for classifying the numerous types of product distributions observable in hydrocracking of pure compounds over bifunctional catalysts with a hydrogenation activity ranging from very weak to very strong (2, 3). The validity of this concept for hydrocracking real feedstocks has also been confirmed, e. g. , by Coonradt and Garwood (4) or by Sullivan and Meyer (5).

Other pure mechanisms of cracking are thermal cracking and hydrogenolysis over metals. Very detailed investigations on platinum catalyzed hydrogenolysis of various alkanes, e. g., the isomeric hexanes, have been published recently (6-8).

Ideal hydrocracking and hydroisomerization of n-dodecane furnished much insight into the primary rearrangement and cleavage reactions of alkylcarbenium ions (9). The system seems to be strongly governed by competitive chemisorption at the acidic sites and, to a lesser degree, at the hydrogenative sites (10). It is the intention of the present paper to extend these data that seem to be of principle significance for the chemistry of catalytic conversion of hydrocarbons, to the lower molecular weight n-alkanes. In literature on hydrocracking some results have been reported concerning reactivities of and product distributions from n-alkanes of different chain length (11, 12). However, as a consequence of the catalysts used, hydrocracking in these cases was far from being ideal.

Experimental

The experiments were carried out in a small flow type fixed bed reactor which has been described in a recent publication (9) along with the methods of analysis by capillary gas-liquid chromatography. Results are reported that were gained with all pure n-alkanes ranging from n-hexane to n-dodecane. Feed hydrocarbons were delivered from Fluka, Buchs, Switzerland (purum). Purity exceeded 99. 5 wt. -% in any case. The Pt/Ca-Y-zeolite catalyst (0. 5 wt. -% Pt, SK 200, Union Carbide, Linde Division; volume of catalyst bed: 2 cm^3; particle size: 0. 2 - 0. 3 mm) was calcined in a dried stream of N_2 and activated in a dried stream of H_2 at atmospheric pressure prior to use. The mass of dry catalyst was 1. 0 g. The total pressure and molar ratio hydrogen : n-alkane were kept constant at 39 bar and 17 : 1, respectively, whereas the reaction temperatures and space velocities were varied.

Conversion of n-Alkanes, Hydroisomerization and Hydrocracking

It was found that hydroisomerization and hydrocracking of n-dodecane over the Pt/Ca-Y-zeolite require low reaction temperatures, a typical value being 275 °C (9). This temperature was chosen in the present work to investigate the influence of chain length on the reactivity of the n-alkanes. In Figure 1 the degree of overall conversion has been plotted versus the superficial

holding time. As expected reactivity of the hydrocarbons increases
with increasing chain length. For a quantitative comparison appro-
ximate values of the initial reaction rates may be derived from
the initial slopes in Figure 1 using the equation

$$- r_a = \frac{1}{100} \cdot \frac{dX}{d\left(\dfrac{m_{cat}}{F_a}\right)}$$

In Table I these values are reported, those for n-nonane and
n-undecane being less accurate on account of the relatively high
degrees of conversion even at low holding times. From n-hexane
to n-undecane a tenfold increase in the initial reaction rate is
observed.

Table I.   Influence of chain length on initial reaction rates
(T = 275 °C)

Feed	$10^2 \cdot (-r_a) \left[\dfrac{\text{mole n-alkane}}{\text{g catalyst} \cdot \text{h}}\right]$
n-C_6H_{14}	0.13
n-C_7H_{16}	0.38
n-C_8H_{18}	0.53
n-C_9H_{20}	1.1
n-$C_{11}H_{24}$	1.3

Hydrogenative conversion of n-alkanes on the Pt/Ca-Y-zeolite
results in two principle reactions, hydroisomerization or hydrocracking, the relative importance of each depending on the reaction conditions. At low severities and correspondingly low degrees
of overall conversion hydroisomerization predominates. With
n-octane at 275 °C, for example, the rate of hydroisomerization
is hardly affected by hydrocracking at low holding times (Figure 2).
At high holding times, however, where the degree of hydroisomerization conversion goes through a maximum the rate of hydrocracking increases. It might be suggested from the shape of the
curves in Figure 2 that hydroisomerization and hydrocracking
are reactions in series.
    A wider insight into the influence of chain length on reactivi-
ties of n-alkanes may be gained by plotting the degrees of hydro-

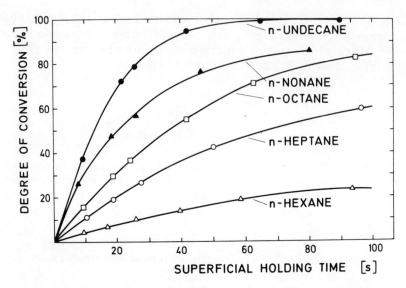

Figure 1. *Hydroisomerization and hydrocracking of* n-*alkanes with different chain length* $(T - 275°C)$

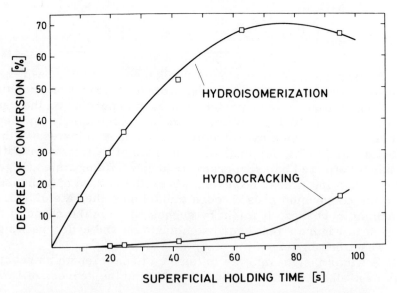

Figure 2. *Hydroisomerization and hydrocracking of* n-*octane at 275°C*

isomerization conversion and hydrocracking conversion versus
reaction temperature (cf. Figures 3 and 4, respectively). Similar
curves are observed for all n-alkanes used. It will be noted that
lower reaction temperatures are required for hydroisomerization
than for hydrocracking. The degrees of hydroisomerization conversion, however, are passing through a maximum which is due
to the consumption of branched isomers by hydrocracking. With
decreasing chain length the position of the maxima shifts towards
higher reaction temperatures reflecting decreasing reactivities.
It should be mentioned that very high maximum values exceeding
60 % are attainable even for long chain n-alkanes like n-decane.
This result has to be attributed to the high hydrogenation activity
of the Pt/Ca-Y-zeolite and is in contrast to hydrocracking over
catalysts of low hydrogenation activity (9, 11, 12) or catalytic
cracking (13), where little or no isomerization of the feed takes
place. With decreasing chain length the height of the maxima
increases, indicating a decreasing tendency for cleavage. An
activation energy of 45 kcal/mole is calculated for the hydroisomerization of, e. g., n-decane.

An even more marked influence of chain length exists for
the hydrocracking reaction (Figure 4). The degrees of hydrocracking conversion rapidly increase with temperature in the
case of n-decane, n-nonane and n-octane. With these hydrocarbons hydrocracking is complete at ca. 300 - 320 °C. In sharp
contrast to this the degree of hydrocracking conversion increases
very slowly with reaction temperature in the case of n-hexane,
whereas a somewhat intermediate behaviour is observed for
n-heptane. It will be shown later in connection with product
distributions that on the Pt/Ca-Y-zeolite hydrocracking of hexane
proceeds via a different mechanism as compared with ideal hydrocracking of the longer chain homologues.

From a practical viewpoint catalysts with high hydrogenation
activity like the Pt/Ca-Y-zeolite provide a high degree of product
flexibility. Long chain feed hydrocarbons in the boiling range of
kerosene may be isomerized with excellent yields which is of
interest for pour point lowering. If, on the other hand, complete
hydrocracking is desired, this may be achieved simply by application of somewhat higher reaction temperatures.

In Figure 5 the generally accepted reaction path (14) for hydroisomerization of n-alkanes has been represented along with different possibilities for the cracking step. The n-alkane molecules
are adsorbed at a dehydrogenation/hydrogenation site where
n-alkenes are formed. After desorption and diffusion to an acidic
site chemisorption yields secondary carbenium ions that rearrange

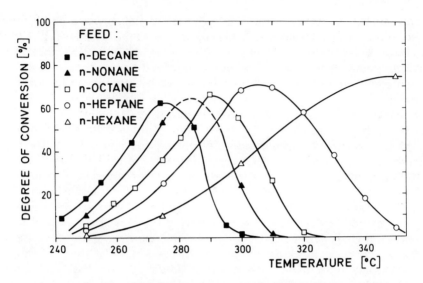

Figure 3.   Influence of reaction temperature on hydroisomerization conversion of n-*alkanes with different chain length* $(F_a = 12 \cdot 10^{-3} \, mole \cdot h^{-1})$

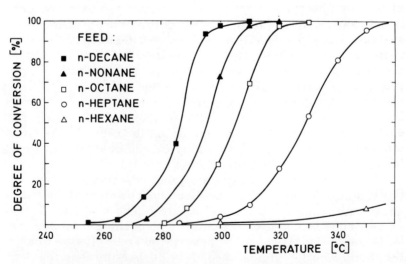

Figure 4.   Influence of reaction temperature on hydrocracking conversion of n-*alkanes with different chain length* $(F_a = 12 \cdot 10^{-3} \, mole \cdot h^{-1})$

into tertiary carbenium ions with a branched carbon skeleton. Desorption of i-alkenes and hydrogenation at a metallic site finally yields i-alkanes. All steps of the overall hydroisomerization reaction are reversible.

Four principle cracking reactions, all of them being irreversible, have to be taken into account: hydrogenolysis of the n-alkane feed; β-scission of straight chain carbenium ions; β-scission of branched carbenium ions; hydrogenolysis of i-alkanes formed by hydroisomerization. It will be shown that the third step (full arrow in Figure 5) represents the main reaction path of ideal hydrocracking. Carbenium ions with a branched carbon skeleton then play the role of key intermediates in that they may either desorb or cleave thus determining the direction of the overall reaction.

While the rate of cleavage is given by temperature, acidity of the catalyst and concentration of i-alkyl cations, the rate of desorption is assumed to be enhanced by the steady state concentrations of n-alkenes, i. e., a high dehydrogenation activity of the catalyst favors hydroisomerization. This is the concept of competitive chemisorption which in ideal bifunctional catalysis keeps the residence times of the alkylcarbenium ions low. Hence, primary products may be obtained which is not the case in catalytic cracking over monofunctional catalysts where formation of carbenium ions occurs by hydride abstraction from the n-alkane rather than via n-alkenes.

According to Figure 5 a series of elementary reactions are involved in hydroisomerization and hydrocracking of n-alkanes. At the time being, the rate controlling step of the overall reaction is unknown because of the lack of a detailed kinetic analysis of the system. Possible interpretations of the influence of chain length upon reactivity are speculative. Maximum concentrations of n-alkenes, limited by thermodynamics, increase with increasing chain length of the feed. The same will be true for rates of adsorption both of the feed and the olefinic intermediates. Rates of surface reactions too may depend on the chain length of the chemisorbed species. The chemistry of ideal hydrocracking will be discussed in terms of detailed product distributions, providing insight into the primary products of bifunctional catalysis.

Hydroisomerization

Literature on hydroisomerization of long chain alkanes $> C_7$ is very limited (2, 9, 15, 16) due to both analytical difficulties and the fact that hydrocracking predominates unless the bifunctional catalyst has been suitably selected. In Figure 6 the composition of the octane and undecane fraction have been plotted versus holding time at a reaction temperature of 275 °C. The following general features of hydroisomerization may be recognized: Mono-branched isomers are primary products from which multiple branched isomers are formed in a consecutive reaction. The chain length of the feed has a marked influence both on rate of hydroisomerization and the amount of multiple branched isomers, which increases considerably with increasing chain length. Qualitatively, this is in agreement with thermodynamics and may be attributed to the number of multiple branched isomers which rapidly grows with increasing chain length.

A more detailed picture of hydroisomerization of n-octane and n-nonane is given in Table II in which product distributions are listed for different degrees of conversion along with thermodynamic equilibrium values. The latter have been calculated from Gibbs free energy data available in literature (17) the accuracy of which, however, is not known. From Table II the following conclusions may be drawn:

Hydroisomerization proceeds towards thermodynamic equilibrium which is approximately reached between the normal, mono-branched and di-branched structures at high degrees of overall conversion. Hydrocracking, however, is severe under these conditions. It is evident from Table II that monomethyl isomers are primary products; the same is apparently true for monoethyl isomers although due to thermodynamic reasons lower concentrations are obtained. Dimethyl isomers including those containing a quarternary carbon atom are formed as secondary products. However, trimethyl isomers are formed very slowly so that their concentrations do not reach equilibrium values. It follows from this that the number of ramifications is deciding as to whether a branched isomer is a primary, secondary or tertiary product in hydroisomerization of n-octane and n-nonane.

Though monomethyl isomers, as a whole, are primary products, the rates of formation of individual members may differ substantially from each other. Furthermore they depend on the degree of conversion. With, e.g., n-decane (Table III) at low degrees of conversion relative rates of formation for 2-methyl-nonane : 3-methylnonane : 4-methylnonane : 5-methylnonane are 1 : 2 : 2 : 1 and shift to 2 : 2 : 2 : 1 at high degrees of conversion.

The latter represent the thermodynamic equilibrium distribution which may easily be understood in terms of statistics: Let m represent the carbon number of the n-alkane feed. If it is even

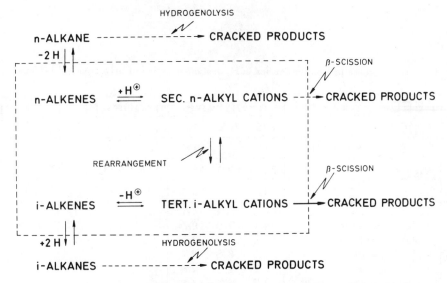

*Figure 5. Reaction scheme for hydroisomerization of n-alkanes on bifunctional cata-
lysts and possible modes of cleavage*

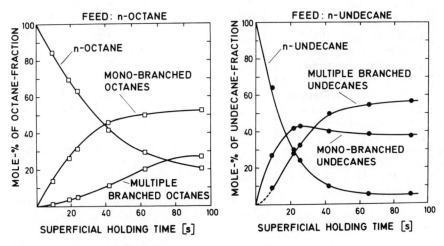

Figure 6. Course of hydroisomerization of n-octane and n-undecane (T − 275°C)

Table II.   Mole-% of Isomers Formed in Hydroisomerization of *n*-Octane and *n*-Nonane

FEED	n- OCTANE			n-NONANE			
T [°C]	275	275	310	275	275	300	327
$F_a \cdot 10^{-3}$ [mole·h^{-1}]	17	3	13	43	7	12	Thermo-
X_{ISO} [%]	29.6	67.0	26.5	25.8	67.1	24.3	dynamic Equili-
X_{cr} [%]	<1	15.6	69.0	<1	9.3	72.6	brium (a)
n-Octane	69.5	20.3	14.0				13.8
2-Methylheptane	9.1	19.0	19.7				14.3
3-Methylheptane	11.4	22.3	23.3				17.3
4-Methylheptane	4.5	8.5	8.7				5.7
3-Ethylhexane	1.5	2.8	2.8				8.0
2,3-Dimethylhexane	0.7	4.1	4.6				2.7
2,4-Dimethylhexane	1.3	8.0	9.2				9.8
2,5-Dimethylhexane	0.7	5.6	6.3				8.5
3,4-Dimethylhexane	0.3	1.2	2.2				3.8
2-Methyl-3-ethylpentane	0.1	0.5	0.7				1.1
2,2-Dimethylhexane	0.5	4.3	4.6				5.0
3,3-Dimethylhexane	0.4	3.2	3.7				4.6
Trimethylpentanes	-	0.2	0.2				5.4
n-Nonane				74.1	25.9	11.1	6.5
2-Methyloctane				5.7	13.7	15.0	9.4
3-Methyloctane				7.8	15.8	16.7	9.9
4-Methyloctane				6.7	13.2	13.5	9.9
3-Ethylheptane				1.4	2.8	3.2	2.6
4-Ethylheptane (b)				1.4	4.0	5.2	5.0
2,3-Dimethylheptane				0.3	2.6	3.8	3.2
2,4-Dimethylheptane				0.4	3.5	5.1	7.9
2,5+3,5-Dimethylheptane				0.9	8.6	11.3	10.9
2,6+4,4-Dimethylheptane				0.4	4.0	6.0	6.4
2,2-Dimethylheptane				0.3	2.3	3.2	5.5
3,3-Dimethylheptane				0.3	1.9	2.7	5.3
Others (c)				0.3	1.7	3.2	17.5

(a) Thermodynamic equilibrium calculated from ref. (17)
(b) Including 3,4-dimethylheptane
(c) Mainly trimethylhexanes

(e. g. , m = 10) then the main chain in the monomethyl isomers contains an odd carbon number (m-1 = 9). In this case a single carbon atom exists for center branching (5-methylnonane) whereas two carbon atoms are available for branching in any other position.

The low initial rate of formation for 2-methylnonane amounting to one half as compared to that of 3-methylnonane or 4-methylnonane reveals a remarkable kinetic feature of hydroisomerization of long chain n-alkanes. It is not observed in hydroisomerization of normal alkanes containing a relatively short chain, e. g. , n-hexane as reported by Bolton and Lanewala (18). More data are required for a detailed understanding of the isomerization of long chain paraffins. Possibly, however, the selectivities found with n-decane at low degrees of conversion reflect the rearrangement chemistry of long chain alkyl carbenium ions.

Table III.   Hydroisomerization of n-decane. Composition of monomethyl isomers (mole-%)
(F_a = 12 · 10^{-3} mole decane · h^{-1})

T [°C]	224	250	274	300
X [%]	2. 3	19	76	99. 8
2-Methylnonane	17. 2	21. 7	27. 8	28. 5
3-Methylnonane	34. 3	32. 3	31. 8	31. 2
4-Methylnonane	31. 2	30. 1	26. 7	26. 8
5-Methylnonane	17. 3	15. 9	13. 7	13. 5

Carbon Number Distribution of Cracked Products

Selectivities of hydrocracking n-alkanes n-$C_m H_{2m+2}$ with even and odd carbon numbers on the Pt/Ca-Y-zeolite are shown in Figures 7 and 8, respectively. Runs have been selected with a medium degree of cracking conversion ranging from 40 to 73 %. In each case the curves are symmetrical indicating pure primary cracking. Methane and ethane as well as C_{m-1} and C_{m-2} are virtually absent. Product distributions like this are typical for ideal hydrocracking.

High hydrogenation/dehydrogenation activity of the bifunctional catalyst is a prerequisite for pure primary cracking of long chain n-alkanes (4, 19). Thus it is the same type of catalyst that

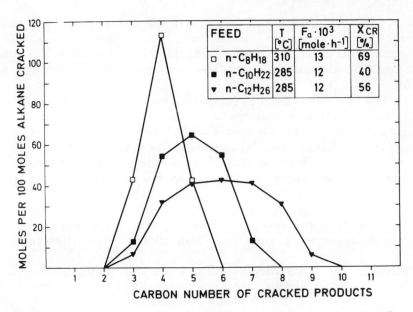

Figure 7.   *Hydrocracking of n-alkanes with an even carbon number. Distribution of the cracked products.*

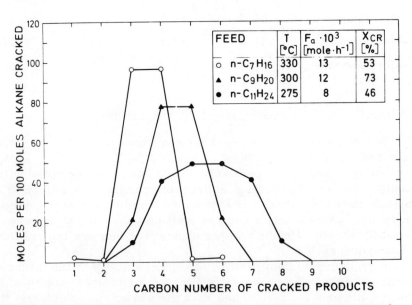

Figure 8.   *Hydrocracking of n-alkanes with an odd carbon number. Distribution of the cracked products.*

makes possible both high degrees of hydroisomerization conversion and pure primary cracking. The mechanistic pathway of
ideal hydrocracking via alkenes and carbenium ions is represented in Figure 5 by a full arrow. The cracked products formed by
β -scission of the long chain carbenium ions may suffer two
different fates: Rapid desorption of b o t h primary fragments
from the acidic sites followed by hydrogenation of the olefinic
intermediates at the metallic sites. Such a pure primary cracking
will result in symmetrical product distributions. If, on the other
hand, desorption of one of the primary fragments from the acidic
site (usually the one with the higher molecular weight) is slow,
secondary cracking may occur. Thus relative rates of desorption
and β -scission are decisive as to whether or not pure primary
cracking may be achieved. The steady state concentration of long
chain alkenes influences the rate of desorption of the primary
fragments (competitive chemisorption) but not the rate of the surface reaction, i. e., β -scission. The concentration of long chain
alkenes, in turn, depends on the dehydrogenation activity of the
bifunctional catalyst.

This picture explains the great differences in product distributions of ideal hydrocracking and catalytic cracking. Furthermore, it is in agreement with the observation that carbon number
distributions are similar in ideal hydrocracking of n-alkanes and
in catalytic cracking of n-alkenes with the same carbon number
(19).

According to the role of competitive chemisorption pure
primary cracking of n-alkanes as shown in Figures 7 and 8 for
medium degrees of cracking conversion may no longer be expected under severe reaction conditions when the feed hydrocarbon
is entirely consumed. Actually, it has been shown (9, 20) that on
the Pt/Ca-Y-zeolite secondary cracking of n-dodecane (m = 12)
or n-decane (m = 10) starts if severities exceed those that are
necessary for a 100 % degree of cracking conversion. The same,
however, is not true if n-alkanes with lower chain length (m $\leq$ 9)
are used. This may be attributed to the low rates of hydrocracking
of all fragments from primary ionic cracking ($\leq$ 6) regardless of
whether competitive chemisorption is effective or not (cf. Figure
4 for hexane). Checks for ideality of a given hydrocracking
system, therefore, should be carried out with an alkane feed of
at least 10 carbon atoms.

Ideal hydrocracking of alkanes requires a minimum of seven
carbon atoms. This may be derived from Figure 9 in which
distributions of the cracked products are compared that were
gained with n-hexane (left hand side) and three of its higher homo-

logues (right hand side) at similar degrees of cracking conversion. Hydrocracking of n-hexane requires a much higher reaction temperature and yields methane and ethane in considerable concentrations. Abstraction of methane is favored over that of ethane. These findings may not be understood in terms of carbenium ion cleavage. Rather, they are in qualitative agreement with the results, that have been reported for hydrogenolysis of the isomeric hexanes catalyzed by platinum on various supports (6-8). From all data of the present work including relative reactivities and distribution of isomers among the cracked products the following conclusions concerning the mechanism of hydrogenative conversion of n-hexane over the Pt/Ca-Y-zeolite are drawn: Hydroisomerization proceeds via the generally valid mechanism outlined in Figure 5. The mechanism of cleavage, on the other hand, differs substantially from the one which is valid for the higher homologues. It is mainly hydrogenolytic starting from both the n-hexane feed and i-hexanes formed by hydroisomerization. Besides, a s m a l l contribution of ionic cleavage of a secondary n-hexylcation, viz. the methyl-n-butylcarbenium ion, β-scission of which yields $C_2 + C_3$, may not be totally ruled out.

In Figure 8 the formation of methane and ethane was shown to accompany ideal hydrocracking of n-heptane to some very small extent. It is somewhat enhanced at low degrees of cracking conversion, which may be seen from the right hand side of Figure 9. Under these conditions superimposition of the methane and ethane abstracting mechanism, viz. hydrogenolysis, may be detected even for n-octane and n-nonane.

Probability of Overall Cracking Reactions

According to the reaction scheme shown in Figure 5 both hydroisomerization and hydrocracking of the n-alkanes (except n-hexane) proceed via b r a n c h e d alkylcarbenium ions. In the range of medium degrees of conversion (40 % $\leq$ X $\leq$ 90 %) both reactions may be investigated simultaneously. A relationship between the products of both types of reaction will be discussed in the present section.

For this purpose selectivities of hydrocracking of the n-alkanes are expressed in terms of probabilities of overall cracking reactions. The latter are calculated from the carbon number distributions of the cracked products (n_{C_i} = number of moles of cracked products with carbon number i, from Figures 7, 8 or 9). Examples for the mode of calculation are represented in the following scheme which is valid for pure primary cracking.

Overall Cracking Reaction	Probability (%)
$C_{10} \rightarrow C_3 + C_7$	n_{C_3} $(= n_{C_7})$
$C_{10} \rightarrow C_4 + C_6$	n_{C_4} $(= n_{C_6})$
$C_{10} \rightarrow C_5 + C_5$	$n_{C_5}/2$
$C_{11} \rightarrow C_3 + C_8$	n_{C_3} $(= n_{C_8})$
$C_{11} \rightarrow C_4 + C_7$	n_{C_4} $(= n_{C_7})$
$C_{11} \rightarrow C_5 + C_6$	n_{C_5} $(= n_{C_6})$

The values defined in this manner do not represent any probability for rupture of definite carbon-carbon bonds in the feed molecule. This term is meaningless if rearrangement of the carbon skeleton precedes the cracking step. Rather, the values indicate the probability of an n-alkane for being hydrocracked according to the overall cracking reaction in question. These probabilities are useful for a comparison with the relative concentrations of the products formed by hydroisomerization (cf. Table IV).

The mechanistic background for such a comparison is illustrated in Figure 10 which represents in more detail the pathway of hydroisomerization and hydrocracking of two n-alkanes. Branched carbenium ions are formed via n-alkenes and linear carbenium ions. Then, either desorption or β-scission may occur in parallel reactions. Desorption (followed by hydrogenation) of a given carbenium ion yields an iso-alkane with the same carbon skeleton. β-scission, on the other hand, yields fragments of definite carbon numbers (β-scissions which would yield C_1 or C_2 are excluded). Thus a comparison between relative concentrations of the iso-alkanes and relative probabilities of the cracking reactions may be informative since both sets of data are determinable independently from each other.

Such a comparison is made in Table IV for all n-alkanes with a chain length ranging from 8 to 12 carbon atoms. Sums have been formed whenever necessary, i.e., if the same set of carbon numbers is formed by β-scission of different carbenium ions, or, if different sets of carbon numbers may be formed by β-scission

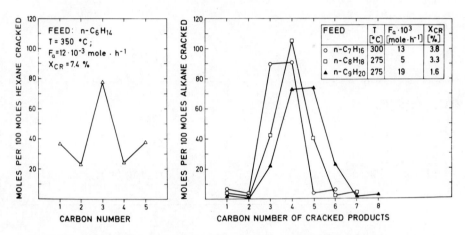

Figure 9. Hydrocracking of n-*hexane and different* n-*alkanes at low degrees of conversion. Distribution of the cracked products.*

FEED	INTERMEDIATES			PRODUCTS
n-HEPTANE $\xrightleftharpoons{-2H}$ n-HEPTENES	$\xrightleftharpoons{+H^{\oplus}}$ NORMAL HEPTYL- CATIONS	$\rightleftharpoons$		2-METHYLHEXANE / $C_4 + C_3$ / 3-METHYLHEXANE
n-DECANE $\xrightleftharpoons{-2H}$ n-DECENES	$\xrightleftharpoons{+H^{\oplus}}$ NORMAL DECYL- CATIONS	$\rightleftharpoons$		2-METHYLNONANE / $C_4 + C_6$ / 3-METHYLNONANE / $C_5 + C_5$ / 4-METHYLNONANE / $C_6 + C_4$ / 5-METHYLNONANE / $C_3 + C_7$

Figure 10. Reaction scheme for hydroisomerization and hydrocracking of n-*alkanes*

Table IV. Hydroisomerization and Hydrocracking of *n*-Alkanes. Comparison Between Relative Concentrations of Iso-alkanes and Probabilities of Overall Cracking Reactions ($F_a = 12 \cdot 10^{-3}$ mole $\cdot$ h^{-1}).

Feed	Reaction	Reaction Temperature [°C]				
		265	275	285	300	310
n-Octane	3-Methylheptane				54.8	54.2
	C$_3$ + C$_5$				44.0	43.0
	2-Methylheptane				45.2	45.8
	C$_4$ + C$_4$				55.0	56.5
n-Nonane	4-Methyloctane		31.2		29.9	26.7
	C$_3$ + C$_6$		21.4		21.7	21.5
	2-Methyloctane } 3-Methyloctane }		68.8		70.1	73.3
	C$_4$ + C$_5$		75.9		77.5	77.9
n-Decane	5-Methylnonane	14.3	13.7	13.4	13.4	
	C$_3$ + C$_7$	14.5	13.5	13.0	13.4	
	2-Methylnonane } 4-Methylnonane }	53.3	54.5	54.7	54.8	
	C$_4$ + C$_6$	53.3	54.5	54.6	54.7	
	3-Methylnonane	32.4	31.8	31.9	31.8	
	C$_5$ + C$_5$	31.1	32.0	32.1	31.4	
n-Undecane	2-Methyldecane } 5-Methyldecane }		48.2			
	C$_3$ + C$_8$ } C$_4$ + C$_7$ }		51.3			
	3-Methyldecane } 4-Methyldecane }		51.8			
	C$_5$ + C$_6$		48.5			
n-Dodecane	3-Methylundecane } 5-Methylundecane }	49.0	47.6	46.9		
	C$_3$ + C$_9$ } C$_5$ + C$_7$ }	48.7	47.8	47.5		
	2-Methylundecane } 6-Methylundecane }	29.1	31.2	32.7		
	C$_4$ + C$_8$	29.6	30.5	31.2		
	4-Methylundecane	21.9	21.2	20.4		
	C$_6$ + C$_6$	21.7	21.7	21.3		

of a particular carbenium ion (e. g., β -scission of both the
3-methyldecyl cation and the 4-methyldecyl cation yields C_5 + C_6;
β -scission of the 5-methyldecyl cation may yield either C_3 +
C_8 or C_4 + C_7).

A good agreement between both sets of data is found for the
long chain n-alkanes, especially for decane and dodecane. In
particular, the great differences in probabilities of overall
cracking reactions from these n-alkanes are reflected by the
relative concentrations of the monomethyl compounds formed by
hydroisomerization. As a whole, the results shown in Table IV
for n-decane, n-undecane and n-dodecane are consistent with
a branching step occuring prior to the cracking step (cf. Figure
10). For n-nonane the agreement between both sets of dat a is
only qualitative whereas principle deviations are observed for
n-octane. Possible interpretations include differences in rate
constants of isomeric carbenium ions either for β -scission or
desorption, or superimposition of a different mode of cracking,
the extent and selectivity of which are unknown.

While there is very little doubt that cleavage in ideal hydrocracking of n-alkanes starts from branched carbenium ions the
exact mode of β -scission may not be derived unambiguously.
Three alternatives are represented in Figure 11 which all provide β -scission of carbenium ions with a single ramification.
Additional modes of cleavage starting from multiple branched carbenium ions might be considered as emphasized by Poutsma (22).
β- Scission as shown by mode No. 1 results in the formation
of a primary cation which would stabilize in a rapid hydride shift.
The occurence of the primary cation may be avoided by assuming
(mode No. 2) the stabilizing hydride shift to be c o n c e r t e d
rather than in series with β -scission. In principle, then, mode
No. 1 and 2 are similar. A third mode of cleavage which has been
taken into consideration (2) starts from secondary carbenium ions
bearing the charge at carbon atom 3 with respect to the tertiary
carbon atom. A concerted hydride shift again avoids the inter-
mediacy of a primary carbenium ion. Although the equilibrium
concentration of the secondary cations required for route No. 3
will be very low, this mode of β -scission is energetically
attractive since a secondary $\longrightarrow$ tertiary reaction is involved.

Starting from the carbon skeleton of 2-methylnonane the
carbon numbers of the fragments are C_4 + C_6, irrespective of
the exact mode of β -scission. Thus each reaction shown in
Figure 11 fits the overall cracking reaction $C_{10} \longrightarrow C_4$ + C_6 and
hence, any of the above modes of cleavage might stand for the
cracking reactions indicated in Figure 10. However, modes

No. 2 and 3 seem to be more appropriate to explain the finding
that n-hexane is excluded from ideal hydrocracking.

Degree of Branching in the Cracked Products

Further insight into the chemistry of hydrocracking may be
gained by examination of the isomer distribution among the
cracked products. Any mode of β-scission shown in Figure 11
predicts both branched and unbranched fragments to be primary
products of cleavage, with a content of branched isomers amounting to ca. 50 %. Actually, higher values are found in ideal
hydrocracking indicating the contribution of β-scission of multiple branched carbenium ions or some secondary isomerization,
i. e., rearrangement reactions following the cracking step. However, the degrees of branching in ideal hydrocracking are lower
than those encountered in hydrocracking over catalysts of low
dehydrogenation activity (12) and they are much lower as compared
with catalytic cracking over monofunctional catalysts (13, 19).
This is consistent with a most effective role of competitive chemisorption in ideal hydrocracking which minimizes the amount
of secondary isomerization.

The influence of chain length of the feed on the degree of
branching of the cracked products will be discussed firstly for a
medium degree of conversion. For this purpose experimental
data have been interpolated to a 50 % degree of cracking conversion. In Figure 12 the content of branched isomers in the alkane
fractions formed by ionic hydrocracking has been plotted versus
chain length of the feed.

Roughly, the content of branched isomers in a given fraction
of the cracked products (e. g., i-butane in the C_4-fraction) is
independent of the chain length of the feed. This is no longer true,
however, if the carbon number fraction in question is formed by
splitting of C_3. Rather, it may be seen from Figure 12 that, if
n-C_mH_{2m+2} represents the feed, the content of branched isomers is extraordinarily high in the fraction C_pH_{2p+2} (p = m-3).

This result is in agreement with any mode of β-scission
represented in Figure 11, since a C_3-fragment formed may not
contain the original ramification. This is illustrated in more
detail in Figure 13 which represents the primary products expected for β-scission according to mode No. 2 and 3. Cracking
of C_{10} has been chosen as an example and again, cleavage of
multiple branched carbenium ions has been omitted.

It may be recognized by summing up the products that both
normal and branched structures are predicted in the C_4-, C_5-

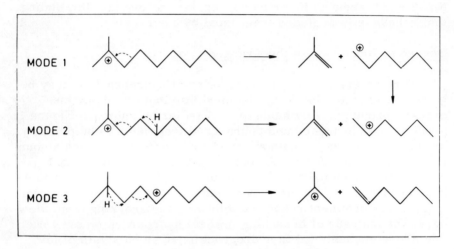

Figure 11.   Modes of β-scission of mono-branched decyl   cations

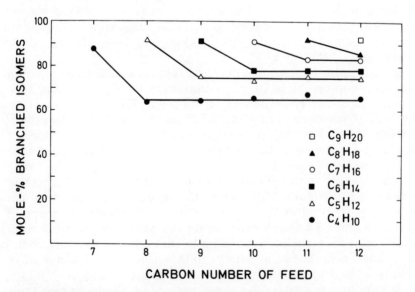

Figure 12.   Hydrocracking of n-alkanes ($X_{cr} = 50\%$). Influence of chain length
of the feed on content of branched isomers in the cracked products.

and C_6-fractions. Both modes of cleavage, however, predict a branched skeleton only for the C_7-fraction. Thus, the content of branched isomers in the fraction resulting by abstraction of C_3 should be 100 %. Actually, this value is approximated, the small deviation of ca. 10 % being due to either secondary isomerization of the branched fragments in direction to unbranched structures or to some small contribution of a non-branching mechanism.

Generally, the degree of branching in an alkane fraction C_pH_{2p+2} formed by hydrocracking of an n-alkane feed $n\text{-}C_mH_{2m+2}$ at a 50 % degree of cracking conversion depends on the value of p as compared with m. It follows from Table V that the alkane fractions C_pH_{2p+2} may be divided into three groups according to their degree of branching.

$3 < p < m-3$	The content of branched isomers depends mainly on p. It increases with increasing p, however, as a good approximation, it is not dependent on m.
$p = m-3$	Degrees of branching are extraordinarily high.
$m-3 < p < m$	Represents the fractions formed by abstraction of C_1 and C_2. Contents of branched isomers are lower as compared with other fractions. They are higher for $p = m-1$ than for $p = m-2$. It is recalled however, that both fractions occur at very low concentrations except when n-hexane is the feed. Thus the degree of branching in these fractions is of interest only for the mechanism of abstraction of methane and ethane rather than for product quality.

Hydrogenolysis has been concluded in a preceeding section to be mainly responsible for the formation of $C_1 + C_{m-1}$ and $C_2 + C_{m-2}$. Branching in the fractions $p = m-1$ and $p = m-2$, then, may be due to hydroisomerization of either the feed or the cracked products according to the reaction sequences:

$$n\text{-}C_mH_{2m+2} \;>\; i\text{-}C_mH_{2m+2} \xrightarrow{(H_2)} i\text{-}C_pH_{2p+2} + C_1 \text{ or } C_2$$

$$n\text{-}C_mH_{2m+2} \xrightarrow{(H_2)} C_1 \text{ or } C_2 + n\text{-}C_pH_{2p+2} \longrightarrow i\text{-}C_pH_{2p+2}$$

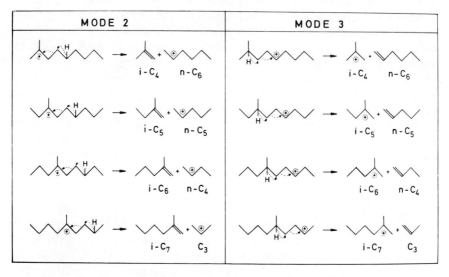

Figure 13.   Primary products in β-scission of mono-branched decyl cations

Table V.   Hydrocracking of *n*-Alkanes ($X_{cr} = 50\%$). Classification of
Cracked Products According to the Degree of Branching.

FEED	MOLE-% OF BRANCHED ISOMERS C_pH_{2p+2}					
$n\text{-}C_m H_{2m+2}$	C_4H_{10}	C_5H_{12}	C_6H_{14}	C_7H_{16}	C_8H_{18}	C_9H_{20}
n-HEXANE [*]	41.0	56.5				
n-HEPTANE	87.5	56	71			
n-OCTANE	63.5	91.5	62	76		
n-NONANE	63.7	75.8	91.0	71	79	
n-DECANE	65.7	73.0	78.0	91.0	73	
n-UNDECANE	67.3	75.7	78.1	83.1	92.0	
n-DODECANE	65.6	74.1	78.0	82.8	85.5	92

[*] $X_{CR} = 7\%$

m-3 < p < m

3 < p < m-3   p = m-3

The higher contents of branched isomers found in the fraction
p = m-1 as compared to the fraction p = m-2 are best explained
by the second route: Secondary hydroisomerization of
n-C_pH_{2p+2} is expected to be faster for p = m-1 due to both
higher reactivity (cf. Figure 3) and competitive chemisorption.

The degree of branching in the cracked products has been
discussed, so far, for a medium degree of cracking conversion.
In Figure 14 hydrocracking of n-heptane and n-octane have been
selected to demonstrate the influence of the reaction temperature
and, hence, of the degree of cracking conversion. Again, a classification of the carbon number fractions is useful:

m-3 < p < m (C_5H_{12} and C_6H_{14} from n-heptane or C_6H_{14}
and C_7H_{16} from n-octane) The degree of
cracking conversion has a pronounced influence on the content of branched isomers in these
fractions. The shape of the curves is similar
to the one that is observed for hydroisomerization of the feed (cf. dashed line for heptane).
This influence of the degree of conversion is
in agreement with the hydrogenolytic mechanism outlined above.

p = m-3 (C_4H_{10} from n-heptane or C_5H_{12} from
n-octane) The degree of branching in these
fractions passes through a maximum at
medium degrees of conversion. The decrease
at high severities indicates an isomerization
reaction i-C_pH_{2p+2} $\longrightarrow$ n-C_pH_{2p+2}
which, due to competitive chemisorption , is
inhibited as long as the system contains high
concentrations of normal- or iso-C_mH_{2m+2}.

The lower contents of branched isomers
found at low degrees of conversion indicate
the contribution of a different type of cracking
mechanism which yields unbranched fragments. In principle, this might be a special
mode of β-scission or, more likely, hydrogenolysis of the unbranched feed. The latter
is in agreement with the shifts in carbon
number distributions encountered at low
degrees of cracking conversion (cf. Figure 9).

3 < p < m-3 (C_4H_{10} from n-octane) These are the most
important fractions in ideal hydrocracking.
It follows from Figure 14, that the content

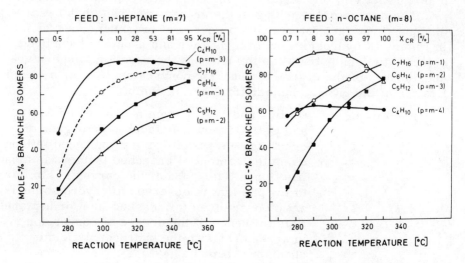

Figure 14.   Hydrocracking of n-heptane and n-octane. Influence of reaction temperature and degree of cracking conversion on content of branched isomers.

of branched isomers depends only slightly on
the degree of cracking conversion. This has
been shown in recent publications (2, 9, 21)
to be generally valid for the fractions in
question. Additional informations concerning distributions of individual isomers,
e. g., a kinetic preference of 2-methyl isomers among the cracked products or the
occurrence of equilib rium concentrations of
quarternary structures at high severities,
have been discussed there.

Conclusions

Low temperature hydrocracking of n-alkanes over suitably
selected bifunctional catalysts with a high hydrogenation/dehydro-
genation activity may be considered to be a pure type of cracking
reaction the products of which differ substantially from those
obtained in catalytic cracking, hydrogenolysis or thermal cracking.
It is accompanied by hydroisomerization and the distribution of
the cracked products is that of a pure primary cracking. Probabilities of overall cracking reactions, as well as branching of the
cracked products, indicate a rearrangement step prior to the
cracking step. Competitive chemisorption plays an important role
by enhancing the rates of desorption of carbenium ions. Thus
in ideal hydrocracking primary products of ionic rearrangement
and cracking reactions are obtainable.

The chain length of n-alkanes has a marked influence on
reactivities for hydroisomerization, and especially for hydrocracking. To a very small extent a methane and ethane abstracting mechanism, probably hydrogenolysis as predicted in a basic
work on bifunctional catalysis (14), is found to be superimposed
when lower carbon number feeds (C_7, C_8, C_9) are used. n-Hexane
is excluded from ideal hydrocracking. On the Pt/Ca-Y-zeolite
catalyst it is cracked via a mechanism that is mainly hydrogenolytic.

From a practical viewpoint ideal hydrocracking offers an
utmost degree of product flexibility. However, as a consequence
of the catalysts required contents of poisons in the feed materials
such as sulfur and nitrogen compounds are strongly limited.

Acknowledgement

The author is indebted to Prof. K. Hedden who generously
supported this investigation. I would also like to thank

Mrs. H. Schlifkowitz and Mr. G. Walter for their excellent experimental work.

Symbols Used

F_a = Feed rate of n-alkane $\left[\text{mole} \cdot \text{h}^{-1}\right]$

m = Carbon number of n-alkane feed

m_{cat} = Mass of catalyst $[\text{g}]$

n_{C_i} = Number of moles of cracked products with carbon number i [moles per 100 moles n-alkane cracked]

p = Carbon number of alkanes formed by hydrocracking

r_a = Reaction rate of n-alkanes $\left[\dfrac{\text{mole n-alkane}}{\text{g catalyst} \cdot \text{h}}\right]$

T = Reaction temperature $[^\circ\text{C}]$

X = Degree of overall conversion $[\%]$

X_{cr} = Degree of hydrocracking conversion $[\%]$

X_{iso} = Degree of hydroisomerization conversion $[\%]$

Literature Cited

(1) Langlois, G. E., and Sullivan, R. F., Preprints, Div. Petr. Chem., A.C.S. (1969), 14 (1), D-18.

(2) Pichler, H., Schulz, H., Reitemeyer, H.O., and Weitkamp, J., Erdöl, Kohle-Erdgas-Petrochem. (1972), 25, 494.

(3) Hedden, K., and Weitkamp, J., Chemie-Ing.-Techn. (1975), 47, 505.

(4) Coonradt, H.L., and Garwood, W.E., Preprints, Div. Petr. Chem., A.C.S. (1967), 12 (4), B-47.

(5) Sullivan, R.F., and Meyer, J A., Preprints, Div. Petr. Chem., A.C.S. (1975), 20, 508.

(6) Anderson, J.R., Advances in Catalysis, (1973), 23, 1.

(7) Matsumoto, H., Saito, Y., and Yoneda, Y., J. Catal. (1970), 19, 101.

(8)   Matsumoto, H. , Saito, Y. , and Yoneda, Y. , J. Catal.
       (1971), 22, 182.

(9)   Schulz, H. , and Weitkamp, J. , Ind. Eng. Chem. , Prod.
       Res. Develop. (1972), 11, 46.

(10)   Weitkamp, J. , and Schulz, H. , J. Catal. , (1973), 29, 361.

(11)   Flinn, R. A. , Larson, D. A. , and Beuther, H. , Ind. Eng.
       Chem. , (1960), 52, 153.

(12)   Archibald, R. C. , Greensfelder, B. S. , Holzman, G. , and
       Rowe, D. H. , Ind. Eng. Chem.(1960), 52, 745.

(13)   Plank, C. J. , Sibbett, D. J. , and Smith, R. B. , Ind. Eng.
       Chem. (1957), 49, 742.

(14)   Weisz, P. B. , Advances in Catalysis (1962), 13, 137.

(15)   Orkin, B. A. , Ind. Eng. Chem. Prod. Res. Develop. (1969),
       8, 155.

(16)   Ciapetta, F. G. , and Hunter, J. B. , Ind. Eng. Chem. (1953),
       45, 155.

(17)   Stull, D. R. , Westrum, E. F. , Jr. , and Sinke, G. C. ,
       "The Chemical Thermodynamics of Organic Compounds",
       John Wiley and Sons, Inc. , New York, 1969.

(18)   Bolton, A. P. , and Lanewala, M. A. , J. Catal. (1970), 18,1.

(19)   Coonradt, H. L. , and Garwood, W. E. , Ind. Eng. Chem.
       Process Des. Develop. (1964), 3, 38.

(20)   Weitkamp, J. , and Hedden, K. , Chemie-Ing. -Techn.
       (1975), 47, 537.

(21)   Schulz, H. , and Weitkamp, J. , Preprints, Div. Petr.
       Chem. , A. C. S. (1972), 17 (4), G-84.

(22)   Rabo, J. A. , "Zeolites Chemistry and Catalysis", A. C. S.
       Chemical Monograph Series, Chapter on "Mechanistic
       Considerations of Hydrocarbon Transformation Catalyzed
       by Zeolites", by M. L. Poutsma, in print.

2

Catalyst Effects on Yields and Product Properties in Hydrocracking

R. F. SULLIVAN and J. A. MEYER

Chevron Research Co., Richmond, Calif. 94802

The flexibility of hydrocracking as a process for refining petroleum has resulted in its phenomenal growth during the past 15 years. Feedstocks that can be converted to lower boiling or more desirable products range from residua to naphthas. Products include such widely diverse materials as gasoline, kerosene, middle distillates, lubricating oils, fuel oils, and various chemicals (1,2).

Commercial hydrocracking is carried out in a single stage or in two or more stages in series. Numerous hydrocracking catalysts have been developed; and the more recent of these have exceptionally long lives, even at severe operating conditions. The choice of the catalyst and of the particular processing scheme will depend on many factors such as feed properties, properties of the desired products, size of the hydrocracking unit, availability of other processing facilities, and various other economic considerations.

In this paper, we will examine several hydrocracking catalysts of varying acidities and hydrogenation-dehydrogenation activities and show differences in product distributions and product properties obtained from two representative domestic feedstocks in the second stage of a two-stage hydrocracking process. In such a process, the feed to the second stage has been hydrofined in the first stage in order to remove most of the impurities such as nitrogen and sulfur.

In particular, we will emphasize (1) total liquid yield, including pentanes and all of the higher boiling product, and (2) octane number of the light product consisting mainly of molecules with carbon numbers of 5 and 6, referred to as $C_5-180°F$ product. High C_5+

liquid yields are desirable in situations in which
butanes and propane are in oversupply or of lower value
than the liquid products. A high octane number in the
C_5-180°F product is particularly desirable because it
is more difficult to upgrade this low boiling fraction
than the higher boiling naphthas, which can be upgraded
relatively easily by catalytic reforming.

Background

Most hydrocracking catalysts of commercial interest are dual functional in nature, consisting of both
a hydrogenation-dehydrogenation component and an acidic
support. The reactions catalyzed by the individual
components are quite different. In specific catalysts,
the relative strengths of the two components can be
varied. The reactions occurring and the products
formed depend critically upon the balance between these
two components.

The acid function of the catalyst is supplied by
the support. Among the supports mentioned in the
literature are silica-alumina, silica-zirconia, silica-
magnesia, alumina-boria, silica-titania, acid-treated
clays, acidic metal phosphates, alumina, and other
such solid acids. The acidic properties of these
amorphous catalysts can be further activated by the
addition of small proportions of acidic halides such as
HF, BF_3, SiF_4, and the like (3). Zeolites such as the
faujasites and mordenites are also important supports
for hydrocracking catalysts (2).

The hydrogenation-dehydrogenation component is a
metal such as cobalt, nickel, tungsten, vanadium,
molybdenum, platinum or palladium, or a combination of
metals. The non-noble metals are usually presulfided
although it has been suggested that the actual hydrogenation activity for the sulfided catalysts exists
in transitory metallic regions on these catalysts (4).
The sulfided metals are generally reported to be less
active for hydrogenation than the noble metal
catalysts (5).

The relationship between the two catalytic components is quite complex. Interactions between the
support and the hydrogenation component can alter this
relationship. For example, Larson et. al. (6) showed
that, with platinum on silica-alumina, a selective
adsorption of platinum by acid sites causes a reduction
in catalyst acidity. Similarly, nickel reacts with the
acid sites on silica-alumina forming nickel salts of
the silica-alumina acid sites. It has been suggested
(7) that one of the effects of sulfiding a nickel on

silica-alumina catalyst is that hydrogen sulfide reacts
with these salts and regenerates the original strong
acid sites of the silica-alumina.

In the reactions of the many component mixtures
that make up most commercial feedstocks, the relationship between the two catalytic components may be further altered by preferential adsorption of certain
hydrocarbon reactants on catalytic sites (8). For
example, polycyclic aromatics have a highly variable
effect dependent on the type and amount as well as
catalyst and reaction conditions. Catalyst poisons
such as sulfur, nitrogen, and oxygen may affect either
or both catalyst components (9).

The literature on the hydrocracking of various
hydrocarbons is summarized in several review articles
(5,10,11) and mechanisms of hydrocarbon reactions
presented. Beuther and Larson (4) and Coonradt and
coworkers (9,12,13,14) compare the reactions of noble
metal catalysts and the non-noble metal sulfides.
Catalysts with high hydrogenation activity such as the
noble metal catalysts are reported to favor the formation of higher boiling products and minimum light
hydrocarbon production. Catalysts with low hydrogenation activity relative to acidity yield product with a
higher ratio of branched paraffins to normal paraffins
and less saturation of aromatics. With the latter
catalysts, the naphtha product has a higher octane
number.

Schutz and Weitkamp (15) show product distributions for the hydrocracking of dodecane on several
noble metals on zeolite catalysts. Product distributions are in general similar to those distributions
previously reported for noble metals on amorphous
supports. These results show no major unexpected
effect of the zeolitic support; differences among the
catalysts tested are related to changes in hydrogenation ability or acidity.

In order to obtain quantitative measurements of
hydrogenation activity and acidity, various schemes
are employed. For example, metal surface area has
been related to hydrogenation activity and the adsorption of bases such as pyridine and ammonia have been
correlated with acidity (6). Some authors have used
certain key reactions involving pure compounds as an
indication of catalytic properties (16). Each of these
methods is useful; however, because of the complex
interdependence of the catalytic functions of the
hydrocracking catalysts and changes in these functions
with catalyst aging, results from each method must be
interpreted with caution.

Coonradt and coworkers (13) use an empirical
hydrogenation activity index based on the aromatic-
naphthene ratio in the hydrocracked product. This
approach does not provide an independent measure of
catalytic properties. However, it has the advantage
that activities are measured under actual hydrocracking
conditions; and changes in catalytic properties with
catalytic aging can be observed.

Most direct comparisons available in the literature among catalysts of varying hydrogenation-to-
acidity ratios are for single pass processing. Because
of the different reactivities of the components of commercial feed mixtures, the specific molecules that
react in a single pass will depend strongly upon the
total conversion. In other words, the most strongly
adsorbed species will react preferentially; and the
resulting product distribution will reflect the properties of these reacting molecules. As the conversion
increases, the properties of the reacting molecules
and therefore the product will change. Also, if two
catalysts with different properties are compared at
constant conversion, the particular species that react
in the presence of one catalyst will not necessarily be
the same species that react preferentially using the
other.

In this paper we compare behavior of catalysts in
extinction recycle hydrocracking. In such a processing
scheme, all of the feed is ultimately converted to
product boiling below a certain predefined temperature.
The reaction paths of individual feed components may
differ from catalyst to catalyst. Product distributions and properties are examined to determine the
general effects of changes in catalytic properties.

Experimental

Equipment. All of the catalysts were tested in
continuous flow, fixed-bed pilot plants equipped for
both liquid and gas recycle operation and continuous
distillation of products. Hydrocarbons boiling above
the desired product end point were recycled to extinction, that is, to 100% conversion of fresh feed. The
product was cooled and passed into a high pressure
phase separator. Here, hydrogen-rich recycle gas was
flashed from the hydrocarbon product and recycled back
to the reactor inlet. Electrolytic hydrogen make-up
was added on demand to maintain constant system
pressure.

The liquid from the high pressure separator was
depressured and flashed at atmospheric pressure. The

liquid hydrocarbon phase was fed to the first of two
continuous distillation columns for the product fractionation. In the first column, the material boiling
above the desired recycle cut point was separated from
the lower boiling product. The overhead from this
column was combined with the gaseous product from the
low pressure flash and fed to the second column in
which the pentanes and lighter gases were separated
from the liquid product. The gaseous product from
this depentanizer was metered and analyzed by gas
chromatography.

Liquid product was distilled batchwise for
determination of liquid yields and product properties.
In the batch distillation, the first liquid product
cut was made at 180°F (true boiling point). Isopentane
and n-pentane were added back to this distillation cut
in the amount measured in the gaseous product. The
resulting blend, mainly consisting of components with
carbon numbers of 5 and 6, is referred to as "C_5-180°F
product."

The boiling ranges of the distillation cuts
(shown as "true boiling point" temperature ranges) were
chosen arbitrarily and do not represent optimum or
maximum potential yields of specific products. In the
examples given, the amount of jet fuel or naphtha
could be altered by changing the boiling ranges,
depending on the desired product properties. Similarly, the recycle cut point could be varied to maximize individual products.

All of the pilot plant tests were made at 60
liquid volume percent per pass conversion below the
recycle cut point. Temperatures were adjusted as
necessary to maintain this conversion. Total pressure
and recycle gas rate were held constant for all of the
runs with a given feed.

Feeds. Properties of two hydrofined test feeds
are given in Table I. The California gas oil blend
was used in tests simulating a hydrocracking unit producing both naphthas and jet fuel, the Mid-Continent
blend in tests representing a unit producing naphtha
as the major product.

Catalysts. Seven experimental catalysts were
prepared with varying hydrogenation activity and
acidity to test the effects of these properties on
product distributions. All of the catalysts were
reasonably stable at test conditions.

In the following table, the catalysts are listed
in order of decreasing ratios of hydrogenation

TABLE I

FEED PROPERTIES

	Hydrotreated California Gas Oil Blend	Hydrotreated Mid-Continent Gas Oil Blend
Gravity, °API	33.9	33.3
Aniline Point, °F	192.7	177.6
Sulfur, ppm	1	9
Total Nitrogen, ppm	0.1	0.2
Mass Spectrometric Type Analysis, LV %		
Paraffins	31.3	19.4
Naphthenes	56.3	63.9
Aromatics	12.4	16.6
ASTM D 1160 Distillation, °F		
St/5	553/589	325/441
10/30	595/617	515/616
50	646	666
70/90	684/732	705/762
95/EP	763/859	802/826

activity to acid activity. Nominally, Catalyst A has
the highest hydrogenation activity relative to its
acidity; Catalyst G has the lowest hydrogenation
activity relative to its acidity. The assignment is
based on a variety of laboratory tests. As indicated
previously, all such tests are not completely unambi-
guous; therefore, the assignment may be regarded as
somewhat arbitrary.

Experimental Catalysts

Catalyst Identification	Hydrogenation Component	Metals Content, Wt %	Support Material
A	Pd	0.5	Activated Clay (Low Acidity)
B	Pd	1.0	Amorphous Silica-Alumina
C	Pd	0.2	Amorphous Silica-Alumina
D	Pd	0.5	Activated Clay (Moderate Acidity)
E	Pd	0.5	Faujasite
F	Pd	0.5	Amorphous Silica-Alumina (Activated)
G	Sulfided Ni	10.0	Amorphous Silica-Alumina

Results With California Gas Oil

Product Distributions. All of the tests with
California gas oil were made at a recycle cut point of
550°F, arbitrarily chosen and not necessarily an
optimum cut point for operation with any of the
catalysts.

Table II compares product distributions for
catalysts near to the extremes of hydrogenation-to-
acidity ratios studied. Catalyst B gives a higher
yield of liquid product including the pentanes and all
higher boiling material (referred to as "C_5-550°F

TABLE II

YIELDS AND PRODUCT PROPERTIES FROM
HYDROCRACKING OF CALIFORNIA GAS OIL
60% PER PASS CONVERSION BELOW 550°F

Catalyst	B		G	
Temperature, °F	598		610	
	Wt %	LV %	Wt %	LV %
No Loss Product Yields **(Based on Fresh Feed)**				
Methane	0.01		0.01	
Ethane	0.02		0.04	
Propane	1.0		1.5	
Isobutane	3.8	5.8	6.3	9.6
n-Butane	1.1	1.6	1.8	2.6
C_5–180°F	10.1	13.2	15.5	20.2
180–280°F	21.6	25.0	25.6	29.5
280–300°F	4.9	5.5	6.4	7.2
300–550°F	59.2	63.9	44.8	48.2
Total C_5–550°F	95.8	107.6	92.3	105.1
Chemical Hydrogen Consumption, SCF/B	950		1000	
Product Properties				
C_5–180°F Product				
Octane Number F-1 Clear	81.5		85.0	
180–280°F Product				
Octane Number F-1 Clear	58.7		63.4	
Type Analysis, LV %				
Paraffins	47.5		42.6	
Naphthenes	52.5		55.5	
Aromatics	0.0		1.9	
280–300°F Product				
Type Analysis, LV %				
Paraffins	44.0		40.1	
Naphthenes	56.0		57.5	
Aromatics	–		2.4	
300–550°F Product				
Type Analysis, LV %				
Paraffins	42.6		41.5	
Naphthenes	57.2		56.0	
Aromatics	0.2		2.5	
ASTM D 86 Dist., °F				
St/5	334/350		333/349	
10/30	358/381		355/375	
50	408		398	
70/90	444/486		436/488	
95/EP	500/531		503/528	

product"). Also, this catalyst produces more material
boiling above 280°F, arbitrarily referred to as jet
fuel. The product is more completely hydrogenated than
that for Catalyst G. Catalyst G, on the other hand,
produces more naphtha, more propane and butane, and
higher octane numbers.

The general trends shown by these product distributions are in basic agreement with the literature
results discussed earlier.

C_5-180°F Product. Table III shows the complete
distribution of hydrocarbons from representative
samples of C_5-180°F product as determined by gas
chromatography for the same two catalysts. Both
product samples contain about 90% paraffins and 10%
cycloparaffins. The C_5-180°F product from Catalyst G
contains a trace of benzene (0.2%); that from Catalyst
B contains no detectable aromatic compound. The small
difference in the total cycloparaffins content shown
in Table III is probably not significant.

The C_5-180°F product consists mainly of C_5 and C_6
paraffins; therefore, it is primarily these components
that determine the octane number.

With all of the catalysts in this study, the ratio
of isopentane to normal pentane varies directly as the
ratio of isohexanes to n-hexane. (In this paper, the
branched C_6 paraffins are collectively referred to as
isohexanes.) Therefore, the iso/normal ratio of the
paraffins of either carbon number can be correlated
with octane number for C_5-180°F product of a given
cycloparaffin content and serve as a convenient indication of catalyst performance.

In the specific example shown in Table III, the
difference in ratios of isoparaffins to normal paraffins in the C_5-180°F product from the two catalysts
results in a four octane number difference.

Figure 1 shows the effect of temperature on
isohexane/n-hexane ratio for Catalysts A-G, inclusive.
Most of the catalysts show a slight decrease in iso-to-
normal ratio with increasing temperature. The ratios
for the group of catalysts tested range from 20/1 to
3/1. The latter ratio is approximately the equilibrium
ratio for singly branched C_6 paraffins to normal
hexane.

Figure 2 is a comparison showing the effect of
temperature on isopentane/n-pentane ratios. The
relationship is quite similar to that shown for the
hexanes.

TABLE III

DISTRIBUTION OF C_5-180°F PRODUCT FROM
HYDROCRACKING OF CALIFORNIA GAS OIL

Catalyst	B	G
Average Catalyst Temperature, °F	598	610
Product, LV % of C_5-180°F		
Butanes	0.4	0.3
Isopentane	41.0	44.1
n-Pentane	9.5	2.8
2,2-Dimethylbutane	0.2	0.04
2,3-Dimethylbutane	2.5	3.2
2-Methylpentane	17.3	19.3
3-Methylpentane	9.3	11.4
n-Hexane	6.3	2.2
Isoheptanes	3.7	4.5
n-Heptane	0.1	0.02
Total Paraffins	90.3	87.8
Cyclopentane	0.5	0.4
Methylcyclopentane	7.1	9.4
Cyclohexane	0.6	0.4
Dimethylcyclopentanes, Ethylcyclopentane	1.3	1.5
Methylcyclohexane	0.1	0.3
Total Cycloparaffins	9.6	12.0
Benzene	–	0.2
Total Aromatics	–	0.2
Octane No., F-1 Clear	81.1	85.0
Isopentane/n-Pentane	4.3	16.4
Isohexanes/n-Hexane	4.6	15.5

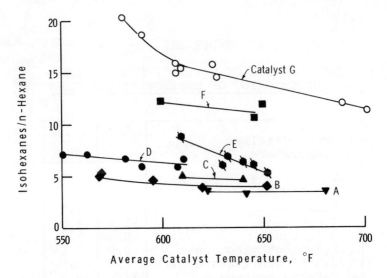

Figure 1. Effect of temperature on isohexanes/n-hexane hydrocracking of California gas oil

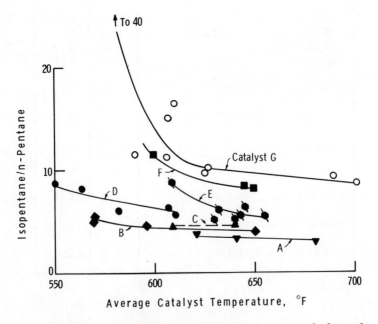

Figure 2. Effect of temperature on isopentane/n-pentane hydrocracking of California gas oil

With both pentanes and hexanes, the iso-to-normal ratios appear to vary inversely with the hydrogenation-to-acidity ratios.

Butanes. The butanes are the principal low boiling products; and, as indicated earlier, catalysts with a low ratio of hydrogenation ability to acidity produce more butane than do those with relatively high hydrogenation activity. However, the differences in isobutane/n-butane ratios among catalysts of widely different hydrogenation-to-acidity ratios are small. Figure 3 shows the effect of temperature on isobutane/ n-butane ratio for Catalysts A, B, F, and G, representing extremes of hydrogenation/acidity ratios. The isobutane/n-butane ratio with Catalyst G is slightly higher than that for the other catalysts at the lower temperatures. However, at higher temperatures, the ratios from all of the catalysts are essentially the same.

This result indicates that there is no direct relationship between iso/normal ratios for the butanes with those for the pentanes and hexanes. The following hypothesis is suggested to explain this result.

An important reaction occurs in hydrocracking which produces isobutane more selectively from cyclo-paraffins than from aromatic compounds. This reaction has been referred to as the paring reaction. It was shown by Egan et al. (17) that cycloparaffins with carbon numbers of 10 or more react very selectively to give isobutane plus lower molecular weight cyclo-paraffins. Alkyl aromatics with side chains of three or more carbons crack mainly by dealkylation, with much less isomerization of the alkyl group (18). Therefore, if the aromatics in the feed are hydro-genated prior to cracking, as would be expected with a catalyst of high hydrogenation activity, a high isobutane/n-butane ratio is favored. With a catalyst of lower hydrogenation activity, cracking would be expected to occur to a greater extent before hydro-genation. Therefore, a lower isobutane/n-butane ratio would be expected.

However, butanes are produced by cracking reactions other than the paring reaction. In reactions such as paraffin cracking, the same factors that favor a high isopentane/n-pentane ratio and a high isopentane/n-pentane ratio and a high isohexanes/ n-hexanes ratio (i.e., high acidity) will favor a high isobutane/n-butane ratio.

Therefore, we have two effects on isobutane/ n-butane ratio that tend to offset each other. The result, with the California gas oil, is that there is

little effect of changing acidity and hydrogenation
activity on the isobutane/n-butane ratio.

This relationship will vary with the aromatic content of the feedstock. With a highly aromatic feed,
the catalyst with the higher hydrogenation activity may
give a higher isobutane/n-butane ratio, although with
the pentanes and hexanes the effect is reversed.

Total Liquid (C_5+) Yield. Product molecules with
carbon numbers of five and higher can be combined in a
fraction referred to as "C_5+" liquid product. This
fraction includes the pentanes and all of the product
boiling below the recycle cut point. The examples in
Table II show that the more acidic catalyst produces
less product in the C_5+ fraction, more butanes and
propane, and less product in the jet fuel boiling range
than does the catalyst with higher hydrogenation
activity.

Figure 4 shows that the total C_5+ yield can be
related to the ratio of the isohexanes/n-hexane. No
results are included for Catalyst A as no analyses were
available in this temperature range.

Within this reasonably narrow temperature span
and with pressure and gas rate constant, the results
correlate quite well. The same general factors which
contribute to a high total liquid yield also result in
a lower iso-to-normal ratio and vice versa.

It should be emphasized that in actual practice it
is the maximum finished product rather than the quantity of C_5+ hydrocrackate that is the most critical
yield affecting economic evaluations of hydrocracking
catalysts. A portion of the hydrocracked naphtha is
usually catalytically reformed to produce high octane
gasoline. Kittrell, Scott, and Langlois (19) presented
and interpreted results showing the optimum relationship between hydrocracking and reforming. Such a
relationship can be used in conjunction with results
of the type described here to evaluate specific hydrocracking catalysts.

Jet Yield. Figure 5 shows the relationship
between the yield of jet (arbitrarily defined as the
product boiling between 280°F and 550°F) and the iso-
to-normal ratio for the hexanes. Although there is
more variation here than shown for the total liquid
yield, this empirical relationship is reasonably good
for most of the catalysts.

Preferential Poisoning of Catalytic Sites. In
order to show that yields and C_5-180°F octane numbers

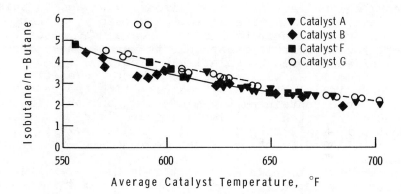

Figure 3. Effect of temperature on isobutane/n-butane hydrocracking of California gas oil

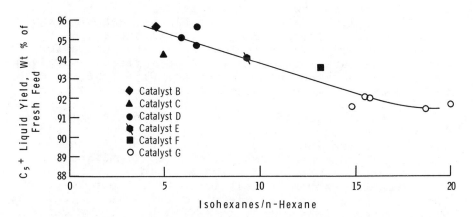

Figure 4. Relationship between liquid yield and isohexanes/n-hexane at 590°-615°F hydrocracking of California gas oil

are directly related to the hydrogenation to acidity
ratio, several experiments were made with Catalyst C in
which catalytic sites were preferentially poisoned.
Nitrogen (as quinoline) was used as a poison for acid
sites; sulfur (as dimethyldisulfide) was used to poison
the metal hydrogenation sites. Admittedly, either
poison may influence both sets of catalytic sites somewhat; however, it is reasonable to assume these secondary effects would be relatively minor.

Table IV shows the results. When the acidity is
preferentially poisoned by equilibrating the catalyst
with a feed containing 8 ppm of nitrogen, the following
effects are noted: (1) Total liquid yield increases by
about 1 wt %; (2) the jet fuel increases by 6 LV %;
(3) the C_5-180°F octane number decreases by two
numbers.

When the hydrogenation activity is preferentially
poisoned by operating with 100 ppm of sulfur in the
feed, (1) the total liquid yield decreases by about 2%;
(2) the jet fuel decreases by 6-7%; (3) the octane
number of the C_5-180°F product increases by more than
three numbers.

Catalyst Aging Effects. The yields for most of
the catalysts remain relatively stable with catalyst
aging. In constant conversion runs such as these, the
catalyst temperature is increased to maintain conversion; and yield stability can be shown by a plot of
yield versus average catalyst temperature. Figure 6
shows results for two amorphous catalysts.

Figure 7 shows the relationship between C_5+ liquid
yield and isohexanes/n-hexane ratio in the temperature
range 640-660°F. The relationship is still good for
most of the catalysts; the average C_5+ yield at a given
isohexanes/n-hexane ratio is perhaps 0.5% lower than
that shown in Figure 4 for the temperature range
590-615°F.

A notable exception to the relationship shown in
Figure 7 is Catalyst E with a faujasite support. In
the temperature span between 640°F and 660°F, the total
liquid yield decreases rapidly without significant
change in the isohexanes/n-hexane ratio.

Figure 8 shows the relationships between temperature, C_5+ yield, and jet yield for Catalyst E. C_5+
yield is relatively stable to 640°F; however, a shift
in product distribution from jet fuel to naphtha occurs
between 610°F and 640°F.

Above 660°F (off scale in the plots in Figures 7
and 8), the C_5+ and jet fuel yield continue to decrease
rapidly. At 677°F the C_5+ liquid yield is 80.4 wt %;
the 280-550°F jet yield is 15.1 LV %.

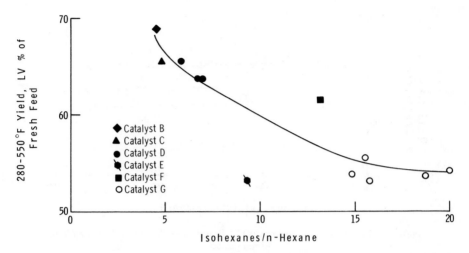

Figure 5.   *Relationship between 280–550°F product and isohexanes/n-hexane at 580°–615°F hydrocracking of California gas oil*

TABLE IV

PRODUCT FROM CATALYST C
WITH CALIFORNIA GAS OIL –
EFFECTS OF SULFUR AND NITROGEN

	Catalyst Temperature, °F	Isohexanes/ n-Hexane	C_5+ Yield, Wt %	280°F+ Yield, LV %	C_5–180°F Octane, F-1 Clear
Before Addition of Sulfur or Nitrogen	609 640	4.9 4.7	94.3 95.2	65.3 64.7	81.0 80.0
8 ppm Nitrogen Added to Feed (Catalyst Equilibrated)	650	3.0	95.8	71.0	78.0
100 ppm Sulfur Added to Feed (After 170 Volumes of Feed Containing Sulfur)	620	12.7	92.7	58.3	84.2

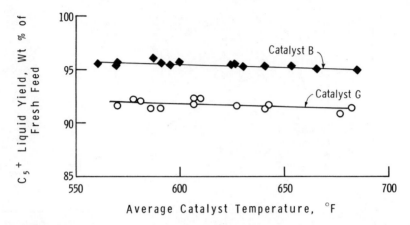

Figure 6.   *Effect of temperature on C_5^+ liquid yield hydrocracking of California gas oil*

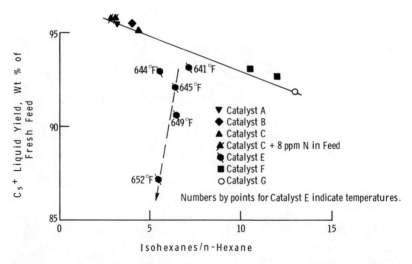

Figure 7.   *Relationship between C_5^+ yield and isohexanes/n-hexane 640°–660°F*

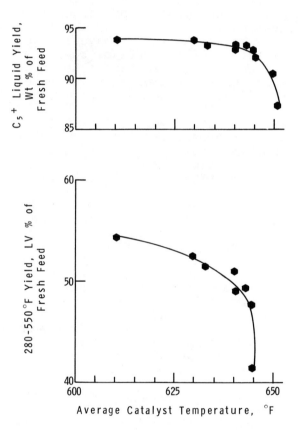

Figure 8. *Effect of catalyst fouling on product composition with catalyst E hydrocracking of California gas oil*

The yield shifts for these faujasite-containing catalysts are accompanied by a buildup of high boiling material (650°F+) in the liquid recycle stream.

The shape selective properties of the aged faujasite catalyst appear to affect the product distribution. Composition of the recycle stream indicates the rate of reaction of the higher boiling molecules is very low. Because high molecular weight molecules did crack more readily in the early part of the run, this result suggests that deposits of carbonaceous material ("coke") cause partial plugging of the catalyst pores. Therefore, the larger molecules have limited access to catalytic sites. The high rate of coking in a narrow temperature range may be related to the hydrogenation-dehydrogenation equilibria of certain polycyclic compounds which, if dehydrogenated, are well known coke precursors.

The rate of catalyst deactivation, as measured by the temperature increase required to maintain conversion, does not increase appreciably during the period that the yield shift occurs. The catalyst remains active for the cracking of relatively low boiling molecules. However, the low yield of product in the jet boiling range indicates that, at the conversion levels used, much initial product from the jet boiling range is further cracked to naphtha and light gases. The secondary cracking of product in the jet range may be the result of several factors: (1) The rates of diffusion of jet product out of the catalyst may be slow due to the partial pore plugging. The longer residence time of the jet product permits much secondary cracking. (2) Partial pore plugging does not allow as many of the higher boiling reactant molecules to reach the catalytic sites. Therefore, there is less competition for catalytic sites; and the apparent rate of reaction of molecules in the jet range may increase.

The catalyst aging effects shown by this faujasite containing catalyst appear to be a general phenomenon; similar effects have been observed in our laboratories with other feedstocks and other zeolite-containing catalysts. Other examples have been reported in the patent literature (20).

This catalyst aging property does not necessarily make such a catalyst unattractive. Rather, it may limit the temperature span in which a desired product distribution can be obtained. With most amorphous catalysts, a run is usually terminated at the point at which catalyst activity and stability are sufficiently low that continuation of the run is no longer attractive.   With certain zeolite-containing catalysts, the product distribution may be the limiting factor.

Results With
Mid-Continent Gas Oil

Catalyst C and Catalyst G were compared for the hydrocracking of the Mid-Continent gas oil at 400°F recycle cut point with naphtha as the major product.

Table I lists the properties of the hydrofined feed; Table V shows yields and product properties at comparable conditions, and Table VI gives detailed chromatographic analyses of representative C_5-180°F products.

The trends are generally the same as those at the higher cut point with California gas oil. The catalyst with the higher ratio of hydrogenation activity to acidity, Catalyst C, produces more C_5+ naphtha; the more highly acidic catalyst, Catalyst G, yields product with a higher octane number.

Conclusions

Most dual functional hydrocracking catalysts exhibit an inverse relationship between C_5+ liquid and the F-1 clear octane number of the C_5-C_6 product. Preferential poisoning of either the acid sites or the hydrogenation-dehydrogenation catalytic sites indicates that both the yields and octanes are related to the ratio of hydrogenation to acidity provided by the catalyst. A high ratio favors high liquid yields; a low ratio favors high octane product.

Some modest changes in the relationship between C_5+ yield and light naphtha octane number occur as catalyst temperature is increased. Also, with certain aged catalysts such as those containing faujasite, changes in catalyst geometry brought about by pore plugging due to carbonaceous deposits may cause substantial deviations from this relationship.

Acknowledgment

The authors wish to thank Messrs. G. E. Langlois and C. J. Egan for their helpful suggestions during the course of this work.

Abstract

Yields and product properties in hydrocracking are influenced by the relationship between catalyst acidity and the hydrogenation-dehydrogenation activity of the

TABLE V

YIELDS AND PRODUCT PROPERTIES FROM
HYDROCRACKING OF MID-CONTINENT GAS OIL
60% PER PASS CONVERSION BELOW 400°F

Catalyst	C		G	
Average Catalyst Temp., °F	634		632	
	Wt %	LV %	Wt %	LV %
No Loss Product Yields (Based on Fresh Feed)				
Methane	0.02		0.03	
Ethane	0.04		0.09	
Propane	1.7		2.4	
Isobutane	7.5	11.4	9.5	14.4
n-Butane	2.1	3.1	2.9	4.2
C_5-180°F	19.1	24.8	21.3	27.7
180-280°F	33.6	39.0	33.2	38.2
280-400°F	38.1	41.9	32.9	35.9
Total C_5-400°F	90.8	105.7	87.4	101.8
Chemical Hydrogen Consumption, SCF/B	1220		1245	
Product Properties				
C_5-180°F				
Octane No., F-1 Clear	81.1		85.7	
180-400°F				
Octane No., F-1 Clear	45.3		55.2	
Type Analysis, LV %				
Paraffins	40.5		37.8	
Naphthenes	59.0		56.1	
Aromatics	0.5		6.1	
ASTM D 86 Distillation, °F				
St/5	220/236		214/228	
10/30	242/264		235/255	
50	288		279	
70/90	318/356		309/350	
95/EP	368/408		365/407	

2. SULLIVAN AND MEYER *Catalyst Effects* 49

TABLE VI

DISTRIBUTION OF C_5-180°F PRODUCT FROM
HYDROCRACKING OF MID-CONTINENT GAS OIL

Catalyst	C	G
Average Catalyst Temperature, °F	627	642
Product, LV % of C_5-180°F		
Butanes	0.8	0.2
Isopentane	36.6	44.6
n-Pentane	8.7	3.5
2,2-Dimethylbutane	0.3	0.1
2,3-Dimethylbutane	2.9	3.8
2-Methylpentane	16.9	19.5
3-Methylpentane	10.1	11.2
n-Hexane	6.6	1.8
Isoheptanes	4.5	3.6
Total Paraffins	87.4	88.3
Cyclopentane	0.5	0.5
Methylcyclopentane	9.1	8.4
Cyclohexane	0.9	0.7
Dimethylcyclopentanes, Ethylcyclopentane	1.7	1.5
Methylcyclohexane	0.4	0.2
Total Cycloparaffins	12.6	11.3
Benzene	–	0.3
Total Aromatics	–	0.3
Octane No., F-1 Clear	81.3	85.7
Isopentane/n-Pentane	4.2	12.7
Isohexanes/n-Hexane	4.6	19.2

dual functional catalysts employed. Hydrocracking
catalysts can be tailored to meet specific refining
objectives. In this paper both amorphous and crystalline catalysts of varying acidity and hydrogenation
activity are examined at constant process conditions
for (1) producing both jet fuel and gasoline and
(2) producing gasoline as the major product. In
general, as the hydrogenation activity of the catalyst
is increased relative to its acidity, total liquid
product (C_5+) yield increases; and the octane number of
the light naphtha (C_5-C_6) decreases. The reverse is
also true. If either the acidity or hydrogenation
activity is preferentially poisoned, the change is
reflected in the yields and the C_5-C_6 octane number.

Literature Cited

1. Scott, J. W., and Kittrell, J. R., Ind. Eng.
    Chem. (1969), 61, 18.
2. Baral, W. J., and Huffman, H. C., World
    Petroleum Congress, Preprint PD 12 (1), (1971).
3. Hansford, R. C., U.S. Patent 3,499,835 (March 10,
    1970).
4. Beuther, H., and Larson, O. A., Ind. Eng. Chem.
    Proc. Design Dev. (1965), 4, 177.
5. Voorhies, A., and Smith, W. M., Chapter in
    "Advances in Petroleum Chemistry and Refining,"
    Vol 8, p 169, Interscience Publishers, New York
    (1964).
6. Larson, O. A., MacIver, D. S., Tobin, H. H., and
    Flinn, R. A., Ind. Eng. Chem. Proc. Design Dev.
    (1962), 1, 300.
7. Langlois, G. E., Sullivan, R. F., and Egan, C. J.,
    J. Phys. Chem. (1966), 70, 3666.
8. Doane, E. P., Preprints, Div. Petroleum Chem.,
    (1967), ACS 12, (4), B-139.
9. Coonradt, H. L., and Garwood, W. E., Ind. Eng.
    Chem. Proc. Design Dev. (1964), 3, 38.
10. Langlois, G. E., and Sullivan, R. F., Advances in
    Chemistry Series No. 97, "Refining Petroleum for
    Chemicals," Chapter 3, p 38, ACS (1970).
11. Beuther, H., McKinley, J. B., and Flinn, R. A.,
    Preprints, Div. Petroleum Chem., (1961), ACS 6,
    (3), A-75.
12. Coonradt, H. L., Ciapetta, F. G., Garwood, W. E.,
    Leaman, W. K., and Maile, J. N., Ind. Eng. Chem.
    (1961), 53, 727.
13. Coonradt, H. L., Leaman, W. K., Maile, J. N.,
    Preprints, Div. Petroleum Chem., (1964), ACS 9,
    (1), 59.

14. Coonradt, H. L., and Garwood, W. E., Preprints, Div. Petroleum Chem., (1967), ACS 12 (4), B-47.
15. Schulz, H., and Weitkamp, J., Ind. Eng. Chem. Prod. Res. Develop., (1972), 11 (1), 46.
16. Myers, C. G., Garwood, W. E., Rope, B. W., Wadlinger, R. L., and Hawthorne, W. P., J. Chem. Eng. Data (1962), 7, 257.
17. Egan, C. J., Langlois, G. E., and White, R. J., J. Am. Chem. Soc., (1962), 84, 1204.
18. Sullivan, R. F., Egan, C. J., and Langlois, G. E., J. Catalysis (1964), 3, 183.
19. Kittrell, J. R., Langlois, G. E., and Scott, J. W., Hydrocarbon Process (1969), 48 (5), 116.
20. Hanson, F. V., and Snyder, P. W., U.S. Patent 3,523,887 (August 11, 1970).

3

Raffinate Hydrocracking with Palladium–Nickel–Containing Synthetic Mica–Montmorillonite Catalysts

JOSEPH P. GIANNETTI and DONALD C. FISHER[a]

Gulf Research and Development Co., P. O. Drawer 2038, Pittsburgh, Penn. 15230

The demand for liquefied petroleum gas (LPG; consisting of propanes and butanes) is projected to increase rapidly in future years.(1) World consumption is dominated by the United States and Japan. Processing of natural gas accounts for the bulk of domestic LPG; however, natural gas production has leveled off forcing the LPG industry to examine other feedstock sources. Japan must look to other countries for future LPG supplies due to environmental and space limitations. An allied problem, especially in the United States, is the continuing need for isobutane to produce valuable alkylates for the gasoline pool.

A recent publication(2) disclosed the use of a palladium-impregnated, nickel-exchanged synthetic mica–montmorillonite catalyst (Pd-Ni-SMM), a 2:1 layer lattice aluminosilicate clay, for the hydrocracking and hydroisomerization of various hydrocarbons including a raffinate fraction to predominately propane and butanes. The results showed that these SMM based catalysts exhibited hydrocracking and hydroisomerization activities greater than could be obtained with Pd-rare earth-zeolite or Pd-H-mordenite. It had earlier been shown that SMM had greater cracking activity than silica–alumina but less than Y zeolite.(3) The greater activity for cracking over the silica–alumina may be due to the increased lability of the hydrogen atoms in SMM.(4)

Our study was undertaken specifically to investigate the processing of a typical raffinate to produce either high yields of LPG or isobutane as well as to determine the octane improvement in the C_5^+ fraction due to hydroisomerization. A 0.7 wt % Pd-15 wt % Ni-SMM catalyst was used for all the experimentation.

Experimental

All units are expressed in both the customary units as well as the "SI" International System of Units. The "SI" designated is shown first followed by the customary units in parenthesis.

(a) Present Address - Alcoa Research Laboratories,
New Kensington, PA 15068

52

Materials. The raffinate used in this study was obtained from a commercial Gulf Oil Corporation refining run and used as received. The composition of this raffinate is shown in Table I.

The 15 wt % Ni-exchanged SMM catalysts were obtained from the Baroid Division of NL Industries. The carbon disulfide was obtained from Fisher Scientific while the palladium tetramine-dinitrate reagent (36.5 wt % palladium) was purchased from Matthey Bishop Corporation. Hydrogen sulfide was obtained from the Matheson Gas Company. All the chemical reagents were used as received. Hydrogen, supplied by the Air Reduction Company, was deoxygenated and dried over Linde 13X molecular sieves.

Catalyst Preparation and Processing Procedure. The 15 wt % Ni-SMM catalyst was impregnated with the palladium tetramine-dinitrate reagent by the incipient wetness technique to result in the incorporation of 0.7 wt % palladium. Since the 15 wt % nickel-exchanged SMM catalyst was obtained from the Baroid Division of NL Industries, no exact detailed synthesis was obtained; however, a typical synthesis from our laboratories has been detailed.(2) All catalysts were either reduced or sulfided prior to use. Approximately 50 cm^3 of catalyst diluted with 50 cm^3 quartz was used for each run. The reduction consisted of 1) treating the catalyst with 3.5 x 10^{-3} kmol H_2/hr (2.9 SCF H_2/hr) at atmospheric pressure while heating to 538°C (1000°F) at a rate of 65.6°C/hr (150°F/hr), 2) holding at 538°C (1000°F) with the hydrogen flow for 1 hour, and 3) cooling to room temperature with a small nitrogen flow. The sulfiding consisted of 1) heating the catalyst to 316°C (600°F) in nitrogen, 2) sulfiding for 3 hours at 316°C (600°F) with 3.6 x 10^{-3} kmol/hr (3 SCF/hr) of a 92% H_2-8% H_2S gas mix, and 3) cooling in nitrogen to room temperature. The Ni-SMM catalysts received contained either about 1 wt % or 0.6 wt % fluoride. Some initial processing comparisons did not show one to be significantly different from the other. Thus, no separation of the runs by fluoride level appeared necessary.

Processing runs were conducted in a fixed-bed automated unit employing a 2.54 cm (1-inch) ID reactor in round-the-clock operations. Operating conditions were 316° to 399°C (600° to 750°F), 7,000 kilopascals: kPa (1,000 lb/sq in gauge: psig), 1.2 to 2 liquid hourly space velocity (LHSV) defined as cm^3 of hydrocarbon feed per cm^3 of catalyst per hour, and 5 hydrogen-to-hydrocarbon mole ratio (to calculate this ratio, the feed was assumed to have a molecular weight of 92). Temperatures were measured by thermocouples extending through the catalyst bed.

Sulfur, in the form of carbon disulfide, was added to the raffinate to give the desired sulfur level. The liquid feed was combined with hydrogen, preheated, and passed downflow through the catalyst. The effluent from the reactor entered a high pressure separator where a hydrogen-rich gas sample was taken. The hydrocarbon product from the separator flowed into a

Table I

RAFFINATE ANALYSIS

Component	Wt %
Isopentane	0.2
n-Pentane	0.2
Hexane Isomers	19.8
n-Hexane	14.6
Methylcyclopentane	4.6
Cyclohexane	0.4
Benzene	1.0
Heptane Isomers	31.5
n-Heptane	9.7
Dimethylcyclopentane	0.5
C_8 and Heavier	17.5

stabilizer where a C_4 and lighter gaseous product and a C_5 and heavier liquid product were taken. The gaseous products were metered and sampled, while the liquid product was collected, weighed, and sampled.

Analyses. Hydrocarbon analyses for the gaseous products were made using a 1200 Varian gas chromatograph with a 30.5 m (100 ft) support coated open tubular squalene capillary and a 21-104 Consolidated Electrodynamics Corp. mass spectrometer. This combination analysis was necessary in order to accurately determine both the hydrogen and hydrocarbon content of the gas as well as a density. The liquid samples only required analyses by the capillary column gas chromatograph.

Micro Research octane numbers were obtained using the standard CFR (Cooperative Fuel Research) knock rating unit.

Results and Discussion

Previous results(2) had shown that a Pd-Ni-SMM catalyst was effective for hydrocracking hexane as well as a raffinate feed. Conclusions showed that this catalyst system when containing two nickel atoms per unit cell (15 wt % nickel) was approximately 15 times more active than a Pd-rare earth-Y zeolite catalyst and 1.2 times more active than Pd-H-mordenite. This same catalyst system (0.7 wt % Pd-15 wt % Ni-SMM) was chosen for our raffinate processing studies.

Comparison of Sulfur-Containing and Sulfur-Free Systems. Our initial experimentation was designed to show the effects of a scheme containing large amounts of sulfur as opposed to a completely sulfur-free system. This experimentation was performed with a sulfided 0.7 wt % Pd-15 wt % Ni-SMM catalyst (hereafter referred to Pd-Ni-SMM catalyst) using a raffinate feed containing 1500 ppm sulfur and a hydrogen reduced Pd-Ni-SMM catalyst with a sulfur-free feed. The LPG yields are presented in Figure 1 with the average product component distributions at $371^{\circ}C$ ($700^{\circ}F$) presented in Table II. These results showed that the sulfur-containing system was more active for hydrocracking but experienced aging especially at the elevated temperatures. The isobutane yield was constant in both systems resulting in significantly higher iC_4/nC_4 and iC_4/LPG volume ratios in the sulfur-free system. Although converting more of the feed, the sulfur-containing system produced significantly less $C_1 + C_2$ than the sulfur-free system.

Various Methods of Incorporating Sulfur. The generally poorer results for LPG from the sulfur-free system indicated that sulfur should be incorporated at least somewhere in the processing system. To determine whether best results are obtained with sulfiding only, sulfur feed addition only, or both, a series

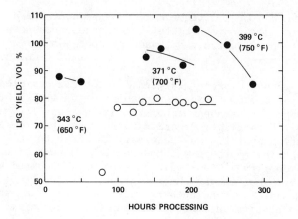

Figure 1. Effect of sulfur in Pd–Ni–SMM raffinate hydrocracking. 7000 kPa (1000 psig), 2 LHSV, 5 hydrogen-to-hydrocarbon mole ratio. Feed—○, raffinate; ●, raffinate + 1500 ppm sulfur. Catalyst—○, reduced 0.7 wt % — 15 wt % Ni–SMM; ●, sulfided 0.7 wt % — 15 wt % Ni–SMM.

Table II

EFFECT OF SULFUR IN Pd-Ni-SMM RAFFINATE HYDROCRACKING

Average Product Distributions at $371°C$ ($700°F$), 7,000 kPa (1,000 psig), 2 LHSV, 5 Hydrogen-to-Hydrocarbon Mole Ratio

	Vol % Yield Based on Feed	
Component	Sulfur Containing	Sulfur Free
Methane + Ethane (Wt %)	3.2	4.7
Propane	46.8	37.5
Isobutane	28.4	28.5
n-Butane	18.9	11.7
C_5 and Heavier	27.0	37.2
C_3/C_4 Ratio	1.0	0.93
iC_4/nC_4 Ratio	1.5	2.4
iC_4/LPG Ratio	0.30	0.37

of catalysts was treated and tested. The results are presented in Table III. The first two runs (A and B) have already been discussed (Figure 1, Table II) and show the effect of a sulfur-free system and a sulfided catalyst along with sulfur doping of the feed. Run C shows that a reduced catalyst with sulfur in the feed (in-situ sulfiding) was generally comparable to a sulfided catalyst with the feed sulfur (Run B with compensation made for aging). Finally, a sulfided catalyst (Run D) with no feed sulfur, while active, produced high C_3 and $C_1 + C_2$ yields. It appears from this that whether the catalyst was reduced or sulfided had little effect on product distribution if sulfur was present in the feed.

Effect of Feed Sulfur Level on Sulfided Catalysts. The importance of sulfur in the feed led to examining varying sulfur levels. Prior to this investigation, all sulfur doping was at the 1500 ppm level. During this investigation an extended run was made in which the sulfur in the feed was incorporated at 0 ppm, 75 ppm, 200 ppm, and finally 500 ppm. The results are presented in Figures 2, 3, and 4. Looking first at Figure 2, it can be seen that the entire product distribution remained constant at a high 108 vol % LPG level through the 200 ppm doping. When the sulfur level increased to 500 ppm, however, pronounced aging began. This aging was accompanied by a decrease in C_3, nC_4, and $C_1 + C_2$ yields, and an increase in iC_4 and C_5^+ yields. Actually, if one assumes that iC_4 is the most valuable hydrocracked product, the distribution became more favorable as the catalyst aged. This is more dramatically shown in Figure 3 where various product ratios are presented. The increase in iC_4 accompanied by decreases in the nC_4 and C_3 yields brought about a more favorable C_3/C_4, iC_4/nC_4 and iC_4/LPG ratios. This indicates that the LPG composition can probably be altered with processing severity. Thus if one wished to produce maximum per pass C_3, high processing severity should be employed while if iC_4 yield is to be maximized, a lower processing severity would be in order.

No mention has yet been made of the C_5^+ material. Since the feed was essentially C_6^+, the C_5's were formed from hydrocracking; however, they have been included in with the C_6^+ material since they would not normally be separated out but would either be recycled with the C_6^+ for further hydrocracking, or, in a once-through operation, be perhaps added with the C_6^+ to the gasoline pool. One thing that occurred to both the C_5's formed and to the remaining C_6's was that there was a high ratio of the branched isomers to their normal counterparts. These results are in keeping with those of Swift and Black(2) who showed that Pd-Ni-SMM catalysts are very effective for hydroisomerization. Figure 4 details the iC_5/nC_5 and iC_6/nC_6 ratios. The iC_5/nC_5 ratio was relatively constant at about 2.4 in the product while the iC_6/nC_6 ratio increased from a feed level of 1.4 to between

Table III

EFFECT OF VARIOUS MEANS OF INCORPORATING SULFUR ON HYDROCRACKING

Conditions -- 371°C (700°F), 7,000 kPa (1,000 psig), 2 LHSV
        5 Hydrogen-to-Hydrocarbon Mole Ratio
Catalyst -- 0.7 Wt % Pd-15 Wt % Ni-SMM

Catalyst Treatment	Wt % Yield $C_1 + C_2$	Volume % Yield				
		C_3	iC_4	nC_4	Total LPG	C_5^+
(A) Reduction (1,000°F), No Sulfur	4.7	37.5	28.5	11.7	77.7	37.2
(B) Sulfiding (600°F), 1500 ppm Sulfur	3.2 (2.6)(a)	46.8 (53)	28.4 (33)	18.9 (20)	94.1 (106)	27.0 (18)
(C) Reduction (1,000°F), 1500 ppm Sulfur	2.2	51.2	31.8	16.8	99.8	22.9
(D) Sulfiding (600°F), No Sulfur	5.2	63.0	26.8	18.0	107.8	14.1

(a) Numbers in parenthesis estimated at initial throughputs prior to aging.

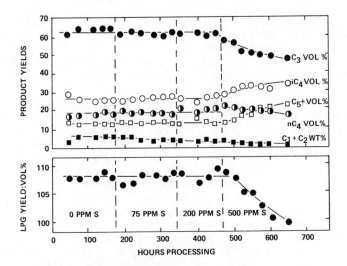

Figure 2. Effect of sulfur level in product yields from hydro-cracking. 371°C (700°F), 7,000 kPa (1000 psig), 2 LHSV, 5 hydrogen-to-hydrocarbon mole ratio. 0.7 wt % Pd — 15 wt % Ni–SMM, sulfided.

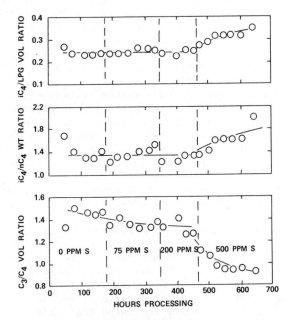

Figure 3. Effect of sulfur level in product ratios from hydrocracking. 371°C (700°F), 7000 kPa (1000 psig), 2 LHSV, 5 hydrogen-to-hydrocarbon mole ratio. 0.7 wt % Pd — 15 wt % Ni–SMM, sulfided.

3 and 5 in the product. The maximum iC_6/nC_6 ratio occurred when the sulfur level was at 75 ppm and 200 ppm sulfur. An increase in the C_5^+ octane number most likely did occur. Additional information on this C_5^+ upgrading is presented later in conjunction with the production of maximum isobutane yields.

Effect of Feed Sulfur Level on a Reduced Catalyst. It has been shown in Table III that generally comparable results were obtained with a sulfur-doped feed and either a reduced or sulfided catalyst. To see if the reduced catalyst had the same response to sulfur level as the sulfided catalyst shown in Figure 2, a run was performed with a 200 ppm and 500 ppm sulfur level. Results are presented in Figure 5 where comparisons are made to the sulfided catalyst. Since the reduced catalyst was not run at 0 and 75 ppm sulfur, the data are lined up at the comparable sulfur levels but not throughputs. The reduced catalyst followed the same pattern as the sulfided catalyst, i.e., no noticeable aging at 200 ppm but aging at 500 ppm. Expectedly, the reduced catalyst became sulfided during the initial processing stages. The indication that the sulfided catalyst was more active should not be considered significant since a 3 to 4 vol % difference in LPG yield could easily be due to slight fluctuations in the processing conditions.

Isobutane Production. The importance of isobutane domestically prompted experimentation aimed at maximizing the isobutane yield. Indications had been obtained from Figures 2 and 3 and discussed previously that as the catalyst aged, isobutane yields increased (at the expense of propane). It appeared likely that processing at less severe conditions would be beneficial towards increasing isobutane yields.

Several runs in the 318°-329°C (605°-625°F), 1.2-2 LHSV range were performed. Results as presented in Figures 6 and 7, show that high yields of isobutane, as high as 50 vol %, can be obtained. In comparing the results from Figure 6 with Figure 2 from the more severe conditions for LPG maximization, the increased yield of isobutane was primarily at the expense of propane with n-butane yields being virtually identical at the two conditions. It is possible that other sets of conditions, for example, 307°C (585°F) and 1 LHSV would produce even higher LPG yields.

In this processing to maximize isobutane yields, a significant amount of C_5^+ normally liquid product remained (e.g. 40 vol % based on feed). The iso-to-normal ratios from the C_5's and C_6's are presented in Figure 8. The same type relationship as previously shown in Figure 4 for LPG maximization resulted, e.g., a high iC_5/nC_5 ratio and a iC_6/nC_6 ratio of 4-5 as compared to 1.4 for the feed. The results indicated that this C_5^+ total fraction should have a significantly higher octane number than the C_6^+ feed. This was indeed the case as the product C_5^+ RON,

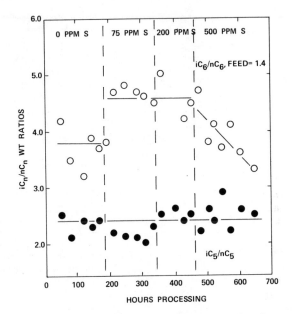

Figure 4. *Effect of sulfur level on isomer distributions. 371°C (700°F), 7000 kPa (1000 psig), 2 LHSV, 5 hydrogen-to-hydrocarbon mole ratio. 0.7 wt % Pd — 15 wt % Ni–SMM, sulfided.*

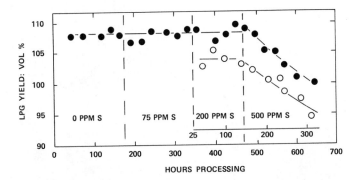

Figure 5. *Effect of sulfur level and catalyst treatment on LPG yields from hydrocracking. 371°C (700°F), 7000 kPa (1000 psig), 2 LHSV, 5 hydrogen-to-hydrocarbon mole ratio. ○, reduced 0.7 wt % Pd — 15 wt % Ni–SMM; ●, sulfided 0.7 wt % Pd — 15 wt % Ni–SMM.*

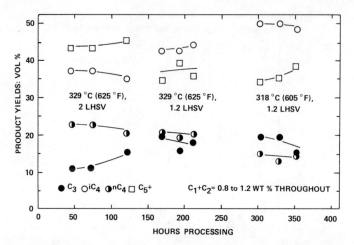

Figure 6.   *Product distribution in hydrocracking for maximum iso-butane yields. 7,000 kPa (1,000 psig), 5 hydrogen-to-hydrocarbon mole ratio, 200 ppm sulfur in feed. 0.7 wt % Pd — 15 wt % Ni–SMM, sulfided.*

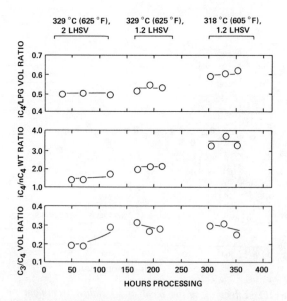

Figure 7.   *Product ratios in hydrocracking for maximum isobutane yields. 7,000 kPa (1,000 psig), 5 hydrogen-to-hydrocarbon mole ratio, 200 ppm sulfur in feed. 0.7 wt % Pd — 15 wt % Ni–SMM, sulfided.*

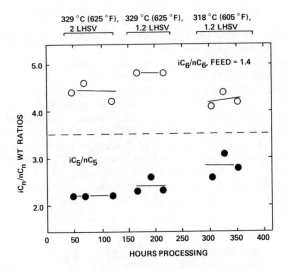

Figure 8. Isomer distributions in products from hydrocracking for maximum isobutane yields. 7,000 kPa (1,000 psig), 5 hydrogen-to-hydrocarbon mole ratio, 200 ppm sulfur in feed. 0.7 wt % Pd — 15 wt % Ni–SMM, sulfided.

clear, was about 80 while the C_6^+ feed was only 54. Thus, a significant upgrading of the normally liquid portion of the product over the feed was obtained. This C_5^+ fraction could have application as a blending component to the gasoline pool. By process manipulations, it would be possible to alter the isobutane-C_5^+ yield octane to meet specified needs.

Acknowledgement

The authors would like to express their appreciation to Dr. Sun W. Chun of Gulf Research & Development Company for his interest and timely suggestions.

Abstract

High yields of LPG or isobutane and octane improvement of the C_5^+ fraction can be simultaneously obtained by hydrocracking raffinate over a palladium-impregnated, nickel-substituted synthetic mica-montmorillonite catalyst (0.7 wt % Pd-15 wt % Ni-SMM). A critical sulfur level of about 100 to 200 ppm in the feed is essential to combine the features of desired product yields and good aging characteristics. Processing conditions ranged from 316° to 339°C (600° to 750°F), 7,000 kilopascals (1,000 psig), 1.2 to 2.0 LHSV and 5 hydrogen-to-hydrocarbon mole ratio with the lower temperatures allowing the high isobutane yields and the higher temperatures maximizing the total LPG yields. No significant aging was observed over 20 days processing as the sulfur level in the feed was increased to 200 ppm.

Literature Cited

(1) Muse, T. P., Hydrocarbon Proc., (1974), 53, 5, 85.
(2) Swift, H.E., and Black, E.R., Ind. Eng. Chem., Product Res. Develop., (1974), 13, 106.
(3) NL Industries, Baroid Division Brochure Introducing Barasym SMM.
(4) Hattori, H., Milliron, D. C., and Hightower, J. W., Division of Petroleum Chemists Inc., Preprints, 165th National Meeting of the American Chemical Society, Dallas, Texas, (April, 1973), Vol 18, pp 33-51.

Hydrocracking Condensed-Ring Aromatics Over Nonacidic Catalysts

WEN-LUNG WU and HENRY W. HAYNES, JR.

Department of Chemical Engineering, University of Mississippi, University, Miss. 38677

In recent years as the shortage of domestic petroleum has become evident, the liquefaction of coal has received serious attention as an alternative source of clean liquid fuels. Research efforts have generally concentrated upon either of two objectives: Earlier investigations were directed toward the production of a "synthetic crude" from coal. This material could then be processed in a (more or less) conventional petroleum refinery to yield a wide range of products from gasoline to heavy fuel oils. More recent efforts have concentrated upon coal liquefaction as a means of producing a "solvent refined coal", i.e. a product low in sulfur content and suitable for use as a fuel to an electric power plant. The principal factor which distinguishes between the synthetic crude and solvent refined coal processes is the amount of hydrogen added during the liquefaction and subsequent processing steps. A synthetic crude must contain somewhere in the neighborhood of 11-14% hydrogen if it is to be processed by conventional petroleum refinery techniques. On the other hand substantial desulfurization can be accomplished if hydrogen is added to the coal to the extent of only one or two per cent to produce a product containing 6-7% hydrogen. The hydrogen generating cost is of course a major factor in the economics of a synthetic fuels plant.

In principle it is possible to produce gasoline and similar products from coal without consuming huge quantities of hydrogen. To illustrate, we can compare the H/C atom ratios for a high volative bituminous coal (0.8) with those for benzene (1.00), toluene (1.14), and mostly paraffinic, petroleum derived gasolines (2.0). Obviously, the more aromatic the refined product, the less is the hydrogen consumption in the overall processing sequence. The problem is that petroleum catalytic cracking and hydrocracking processes require that the feedstocks be highly saturated prior to, or during cracking. More specifically, if one attempts to hydrocrack a multicondensed ring aromatic species, the tendency is to successively hydrogenate and split off terminal end rings to produce ultimately a single aromatic molecule and substantial quantities of C_4 and lighter gases (1, 16, 17, 22). This is true even though hydrogenation at inner ring locations is favored kinetically (18).

The commerical cracking of petroleum hydrocarbons is almost universally executed over catalysts having an acidic component. For many years amorphous silica-alumina based catalysts were the mainstay of the industry. In more recent years cracking activities have been enhanced by orders of magnitude by incorporating crystalline silica-aluminas or molecular sieves (multivalent and hydrogen forms) into the catalyst composition. While acidic catalysts are by far the most active cracking catalysts available, they are not selective towards cracking at inner rings of partially saturated condensed-ring aromatics of the type present in coal liquids. As we discuss below, there are suggestions in the literature that nonacidic catalysts, or catalysts of low acidity might offer advantages in terms of selectivity. The cracking reactions that take place over such catalysts would presumably take place by free radical mechanisms -- similar to the situation encountered in thermal cracking.

A study of the thermal hydrogenolysis of phenanthrene at 80 atm and temperatures of 475°C and 495°C was recently reported by Penninger and Slotboom (11). The main hydrogenation products were 1,2,3,4-tetrahydrophenanthrene and 9,10-dihydrophenanthrene. At the higher temperature cracking products appeared. The principal products were, in decreasing order of abundance, 2-ethylnaphthalene, ethylbiphenyl, 2-methylnaphthalene, naphthalene, biphenyl and diphenylethane. Several reaction paths were suggested by their data. The presence of sizeable quantities of ethylbiphenyl and biphenyl was explained by a mechanism involving the ring opening of dihydrophenanthrene between the 10 and 11 carbon atoms. Thus we see here for the first time evidence of hydrocracking the phenanthrene molecule at the central ring. Still,the major reaction path involved cleavage of end rings.

It can be argued, somewhat naively perhaps, that the main obstacle to hydrocracking 9,10-dihydrophenanthrene and similar molecules, is the 12,13- or "biphenyl" like bond. It is noted that biphenyl itself was observed to be inert to catalytic cracking over an acidic,silica-alumina cracking catalyst (8). On the other hand the hydrogenolysis of biphenyl takes place thermally at temperatures in the neighborhood of 1300°F (10). Gardner and Hutchinson found that low surface area, low acidity, metal oxide catalysts were effective in hydrocracking polyphenyls including biphenyl (7). Acidic cracking catalysts produced much lower conversions and substantial yields of coke. These observations were a major factor in our selection of a chromia-alumina catalyst for this investigation. Chromia-alumina has demonstrated an ability to catalyze many free radical reactions.

Chromia-Alumina Catalysts. Chromia-alumina catalysts are most notable for catalyzing the dehydrogenation and dehydro-

cyclization reactions of hydrocarbons and the polymerization of
olefinic hydrocarbons. Relative to metal dehydrogenation
catalysts such as Fe, Ni, and Co, chromium oxide and chromia-
alumina catalysts possess little activity for activating the
the carbon-carbon bond (14). Chromia-alumina is remarkably
insensitive to poisons which might be present in the starting
material, though water vapor does act as a temporary poison (19).

Chromium oxide by itself is a highly active catalyst, but
this activity is not retained on prolonged use or upon regener-
ation by oxidation. This decay is due to conversion to crystalline Cr_2O_3 which has been shown to be inactive (20). But if
the chromium oxide is deposited on alumina, it retains its
original high activity through regeneration at quite high temperatures (9). The alumina has an additional effect in that it
stabilizes the chromium against reduction below the Cr^{+3} state
in hydrogen atmospheres at temperatures as high as 500°C (21).
Finally the alumina may or may not, depending upon the method of
preparation, introduce an acidic function into the characteristics
of the catalyst. For example, Pines and Csicsery (13) observed
that chromia-alumina catalysts composed of alumina obtained by
hydrolysis of aluminum isopropoxide and alumina prepared from
potassium aluminate behaved quite differently in the dehydrogenation and dehydrocyclization of C_5 and C_6 hydrocarbons. The
former catalyst, designated catalyst A, produced extensive
skeletal isomerization; whereas, the latter, designated catalyst
B, produced olefins with essentially the same skeletal arrange-
ment as the starting paraffins. Catalyst A yielded large
quantities of aromatics -- benzene, toluene and xylene -- from
n-pentane. Formation of these products could be explained by an
acid catalyzed polymerization followed by isomerization and β-
scission to form propylene and isobutylene fragments which could
then undergo an aromatization reaction. Catalyst B produced
only negligible quantities of aromatics.

Pines and his co-workers are responsible for a great body
of literature eludicating the mechanisms of hydrocarbon reactions
over chromia alumina. [For number thirty-eight in the series see
reference (12)]. In general it is observed that reactions over
the "nonacidic" chromia-alumina B can be explained by free radical
mechanisms. In a study of the dehydrocracking of C_6-C_8 paraffins
Csicsery and Pines made several interesting observations (3):
Cracking occurred most predominantly between substituted carbons,
with little cracking of normal paraffins taking place. This
suggests a homolytic splitting of the molecule into free radicals
since the stability of free radicals increases with substitution.
Cracking was very much inhibited if neither of the fragments
could form an olefin. A comparison was made between the
cracking of 3-ethylhexane and 3-ethylhexenes. With the paraffin,
cracking occurred almost exclusively between primary and tertiary carbon atoms. With olefins, cracking took place β to the
double bond suggesting a mechanism involving the allyl free
radical.

   The major products obtained from the dehydrogenation of n-
butylbenzene over chromia-alumina were 1-phenylbutenes (2).
This reaction was accompanied by substantial hydrogenolysis of
the side chain to produce ethylbenzene (or styrene) and ethylene
(or ethane). Dealkylation was a relatively minor reaction.
The product distribution again suggests that free radical intermediates were involved. A review of the physical-chemical
properties of chromia-alumina was presented by Poole and
McIver (15).

Experimental

   The experimental portion of this investigation was conducted
in the steady-flow microreactor illustrated in Figure 1. The
reactor consisted of a tube of 1/2 inch heavy wall (0.083 inch)
type 316 stainless steel heated by a Marshall tubular furnace,
model 1016. Liquid phenanthrene was metered into the reactor by
a precision Ruska proportioning pump, model 2252-BI, with a
heated barrel. Various discharge rates from 2 cc/hr to 240 cc/hr
could be obtained by selecting the proper choice of gear ratios.
The hydrogen flowrate was monitored by a flow meter constructed
of a 34 inch length of 0.009 inch I.D. capillary tubing and a
Barton model 200 differential pressure cell. The capillary
pressure and reactor pressure were controlled respectively by a
Tescom pressure regulator, model 26-1023-002 and a Tescom back
pressure regulator model 26-1723-24. Flow rates were controlled with a Hoke milli-mite needle valve. The high pressure
accumulator was constructed of 3/4 inch schedule 80 stainless
steel pipe with a Grayloc closure.

   High purity hydrogen (99.995% according to manufacturer's
specifications) was supplied by the Matheson Gas Products Company in 3500 psig cylinders. Phenanthrene, 98+% purity, melting
point 99-101°C, was purchased from the Aldrich Chemical Company.
The chromia-alumina catalyst, designated CR-0103-T,was supplied
by the Harshaw Chemical Company in the form of 1/8 inch tablets.
According to the manufacturer, the catalyst contains 12% Cr_2O_3
on activated alumina, the surface area is 63 m^2/gm, and the
pore volume is 0.35 cc/gm.

   Approximately five grams of full size catalyst particles
were charged to the reactor. A preheat zone of glass beads was
situated above the catalyst bed. Thermocouples were installed to
a depth of 1/2 inch into both ends of the catalyst bed. With a
fresh catalyst installed, the system was pressured with hydrogen
and tested for leaks. Hydrogen flow was started and the temperature raised slowly (160°F/hr maximum) to a temperature of
500°F. At this point the liquid feed pump was switched on. The
furnace temperature was gradually brought up to reaction conditions by raising the temperature at a rate not to exceed 160°F/hr.
Prior to beginning a yield period, a period of time sufficient
for three displacements of the reactor was allowed to insure

steady-state operation. The yield period was begun my momen-
tarily cutting off the feed pump, depressuring and draining the
high pressure accumulator, and pressuring the accumulator with
pure hydrogen (through a line not indicated on the simplified
schematic). The exiting line was directed through a low pressure accumulator (in dry ice-acetone), a wet test meter and a
polyethylene gas bag. At the termination of the yield period the
high pressure accumulator was depressured through the wet test
meter and liquid drained. The liquid product was capped and
stored in a freezer. The contents of the gas bag were analyzed
immediately. When it became necessary to place the unit on
over night hold the feed pump was switched off and the reactor
temperature lowered to 500°F. Hydrogen flow was maintained for
the duration of the hold period.

Both the liquid and gas products were analyzed by gas chromatography. The column for the liquid analysis was 20% Apiezon L
on 60-80 mesh Chromosorb P. The column measured 1/4 inch by 7
feet. The gas analysis utilized a 1/4 inch by 10 foot column of
60-80 mesh Chromosorb 102. Temperature programming was required
in both analyses. Identification of the GC peaks was based on
retention time of pure compounds when these were available. In
addition, two of the samples were analyzed by combined gas chro-
matography-mass spectrometry. By comparing the observed mass
spectrometer fragmentation patterns with tabulated patterns it
was possible to identify virtually every component in the product. Further details are available in the theses by Wu (23)
and Early (4).

Results

Sixteen yield periods were successfully completed at the
conditions summarized in Table I. Product yields are tabulated
in Table II. These yields have been adjusted to force a 100%
carbon material balance.

While it was beyond the scope of this investigation to
evaluate catalyst deactivation kinetics, it was deemed necessary
to maintain a record of any declines in catalyst activity. The
selected "Base Conditions" of operation -- 800°F, 2000 psig,
1.0 gm/hr/gm -- were repeated periodically for this purpose. Two
measures of catalyst activity, phenanthrene conversion and conversion to C_{14}, are plotted versus hours on catalyst in Figure 2.
While both indicators reveal that catalyst deactivation is significant, the catalyst does remain active after 90 hours on
stream. No analysis of the spent catalyst was undertaken; however, it was surprisingly observed that the catalyst still
retained its original green color at the run termination. Evi-
dently carbon deposition was not extensive despite the high
temperatures employed. The hydrogen partial pressures experienced by the catalyst are much higher than those experienced in
the usual applications of chromia-alumina, and the extent to

TABLE I

PRODUCT YIELDS

RUN WLW02. HYDROCRACKING OF PHENANTHRENE OVER 5.017 GMS HARSHAW CR-0103-T

Yield period no.	1	2	3	4	5	6	7	8	9	10	11	12	13	14	15	16
Yield period length, hrs.	3.00	3.00	6.00	1.00	1.50	3.00	3.00	3.00	3.00	3.00	3.00	2.75	4.00	1.00	1.00	3.00
Temperature, °F	756	805	805	810	805	800	806	803	854	806	898	949	999	1002	1009	803
Pressure, psig	2000	2000	2000	2000	2000	2000	2500	3000	3000	2000	3000	2800	2800	2800	2600	2000
High press. accumulator temp., °F	250	230	235	255	295	314	275	285	228	210	110	90	88	100	462	260
Liquid feed rate, cc/hr	5.1	5.1	2.5	16.3	8.6	5.2	5.1	5.2	5.0	4.9	4.9	5.1	5.0	16.9	30.7	5.0
Liquid space velocity, gm/hr/gm	1.04	1.04	0.51	3.29	1.74	1.05	1.03	1.05	1.02	1.00	0.99	1.03	1.02	3.42	6.21	1.01
Space time $(gm/hr/gm)^{-1}$	0.97	0.96	1.95	0.30	0.57	0.96	0.97	0.95	0.98	1.00	1.01	0.97	0.99	0.30	0.16	0.99
Approx. H_2 treat rate, L(STP)/hr	6.7	6.7	4.4	14.3	8.9	6.7	6.7	8.9	8.9	6.7	8.9	8.9	8.9	24.0	21.0	6.7
L(STP)/gm	1.29	1.29	1.73	0.86	1.02	1.28	1.30	1.70	1.75	1.34	1.79	1.72	1.75	1.40	0.67	1.33
Exit gas rate, L(STP)/hr	15.1	10.7	4.7	24.9	16.9	6.0	10.6	12.8	16.9	9.2	11.9	20.7	9.6	37.7	25.4	11.7
Cum. hrs. on catalyst	4.5	10.0	20.5	24.5	27.8	32.5	38.8	45.8	51.8	58.2	64.7	69.8	76.4	79.9	81.9	86.8
Cum. gms oil/gm catalyst	4.6	10.1	16.3	21.0	28.6	33.8	40.2	47.2	53.3	59.6	65.7	70.8	78.3	85.3	95.8	103.3
Liq. mat'l balance, wt %	84.3	103.8	103.0	99.6	104.5	90.6	99.0	106.7	104.1	84.9	85.0	83.2	75.2	46.3	97.2	100.8
Carbon mat'l bal., wt %	83.1	102.4	101.4	98.6	103.2	89.3	97.5	105.2	106.4	84.1	83.1	99.1	107.8	87.6	113.4	99.4
CORRECTED YIELDS BASED ON LIQUID FEED																
Feed conversion, mole %	60.1	74.0	86.2	50.2	57.5	69.5	81.1	78.8	87.6	63.5	86.1	83.7	88.2	82.4	49.1	62.2
Conversion to C_{14}^- mole %	2.7	4.9	7.2	3.2	3.2	2.8	3.8	3.6	15.5	2.5	26.3	50.7	74.4	69.6	30.1	4.1
Hydrogen consumption, L(STP)/gm	0.17	0.27	0.36	0.14	0.16	0.24	0.31	0.29	0.43	0.22	0.38	0.59	0.85	1.04	0.38	0.18
wt %	1.5	2.4	3.2	1.3	1.5	2.1	2.8	2.6	3.8	2.0	3.4	5.3	7.7	9.4	3.4	1.7
Gas yield (C_1-C_4), wt %	0.07	1.1	1.6	0.22	0.17	0.6	1.3	1.2	6.0	0.94	1.1	21.3	37.9	56.5	17.7	0.30
Liq. yield (C_5^+), wt %	101.4	101.3	101.6	101.1	101.3	101.5	101.5	101.4	97.8	101.0	102.3	84.0	69.7	52.9	85.7	101.4
C_5-C_8 yield, mole %	0.0	0.0	0.0	0.0	0.0	0.0	0.0	0.0	0.0	0.0	4.1	8.9	22.9	7.0	0.46	0.0
C_9-C_{12} yield, mole %	3.4	5.1	7.3	3.9	3.8	2.9	3.4	3.3	13.2	2.1	30.4	37.8	42.3	24.3	19.3	4.9
C_{13}-C_{14} yield, mole %	97.3	95.1	92.8	96.8	96.8	97.2	96.2	96.4	84.5	97.5	73.7	49.3	25.6	30.4	69.9	95.9

TABLE II   PRODUCT YIELDS

CORRECTED PRODUCT YIELDS BASED ON LIQUID FEED, MOLE

RUN WLW02, HYDROCRACKING OF PHENANTHRENE OVER 5.017 GMS HARSHAW CR-0103-T

Compound	1	2	3	4	5	6	7	8	9	10	11	12	13	14	15	16
C 1H 4 Methane	0.00	0.04	0.03	0.00	0.04	0.36	0.39	0.04	3.12	0.06	0.08	8.10	64.83	88.85	19.92	0.00
C 2H 6 Ethane	0.43	2.26	1.27	0.69	0.53	0.91	1.25	1.51	5.46	1.65	1.69	17.72	70.05	99.98	29.47	0.66
C 3H 8 Propane	0.00	0.61	0.98	0.02	0.04	0.34	0.61	0.52	5.72	0.57	0.65	21.14	48.25	71.56	25.18	0.00
C 4H10 Isobutane	0.00	0.00	0.00	0.00	0.00	0.00	0.00	0.00	0.00	0.00	0.00	0.00	1.28	2.18	0.50	0.00
C 4H10 Butane	0.00	1.61	3.54	0.30	0.21	1.15	2.65	2.44	10.35	1.59	2.07	37.86	24.30	40.53	13.87	0.57
C 5H12	0.00	0.00	0.00	0.00	0.00	0.00	0.00	0.00	0.00	0.00	0.00	0.00	0.17	0.00	0.00	0.00
C 5H12	0.00	0.00	0.00	0.00	0.00	0.00	0.00	0.00	0.00	0.00	0.00	0.88	0.54	0.10	0.00	0.00
C 5H10 Cyclopentane	0.00	0.00	0.00	0.00	0.00	0.00	0.00	0.00	0.00	0.00	0.00	0.33	0.56	0.15	0.00	0.00
C 6H14	0.00	0.00	0.00	0.00	0.00	0.00	0.00	0.00	0.00	0.00	0.00	0.28	0.30	0.08	0.00	0.00
C 6H14	0.00	0.00	0.00	0.00	0.00	0.00	0.00	0.00	0.00	0.00	0.00	0.42	0.65	0.00	0.00	0.00
C 6H 6 Benzene	0.00	0.00	0.00	0.00	0.00	0.00	0.00	0.00	0.00	0.00	2.58	1.68	7.06	2.45	0.00	0.00
C 7H16	0.00	0.00	0.00	0.00	0.00	0.00	0.00	0.00	0.00	0.00	0.36	0.45	0.49	0.32	0.00	0.00
C 7H16	0.00	0.00	0.00	0.00	0.00	0.00	0.00	0.00	0.00	0.00	0.24	0.36	0.93	0.25	0.00	0.00
C 7H16	0.00	0.00	0.00	0.00	0.00	0.00	0.00	0.00	0.00	0.00	0.69	2.33	7.05	2.18	0.00	0.00
C 7H 8 Toluene	0.00	0.00	0.00	0.00	0.00	0.00	0.00	0.00	0.00	0.00	0.00	0.37	0.29	0.00	0.00	0.00
C 7H16	0.00	0.00	0.00	0.00	0.00	0.00	0.00	0.00	0.00	0.00	0.25	1.55	3.77	1.37	0.00	0.00
C 8H10 Ethylbenzene	0.00	0.00	0.00	0.00	0.00	0.00	0.00	0.00	0.00	0.00	0.23	0.23	0.55	0.12	0.46	0.00
C 8H10 Xylenes	0.00	0.00	0.00	0.00	0.00	0.00	0.00	0.00	0.00	0.00	0.00	0.00	0.59	0.00	0.00	0.00
C 8H10	0.00	0.00	0.00	0.00	0.00	0.00	0.00	0.00	0.00	0.00	0.00	0.00	0.29	0.10	0.00	0.00
C 8H18	0.00	0.00	0.00	0.00	0.00	0.00	0.00	0.00	0.00	0.00	0.00	0.00	0.00	0.00	0.00	0.00
C 9H12 Alkylbenzenes	0.00	0.00	0.00	0.00	0.00	0.00	0.00	0.00	0.00	0.00	0.00	0.00	0.00	0.00	0.00	0.00
C 9H12 Alkylbenzenes	0.00	0.00	0.00	0.00	0.00	0.00	0.00	0.00	0.00	0.00	0.00	0.00	0.00	0.30	0.00	0.00
C10H14 Alkylbenzenes	0.00	0.00	0.00	0.00	0.00	0.00	0.00	0.00	0.00	0.00	0.00	0.36	0.00	0.32	0.89	0.00
C10H14 N-Butylbenzene	0.00	0.00	0.00	0.00	0.00	0.00	0.00	0.00	0.00	0.00	2.42	4.83	5.38	2.03	0.95	0.00
C10H18 Decalin (Trans)	0.00	0.00	0.00	0.00	0.00	0.00	0.00	0.00	0.58	0.00	1.17	2.56	3.66	1.52	0.00	0.00
C11H16 Methylbutylbenzenes	0.00	0.00	0.00	0.00	0.00	0.00	0.00	0.00	0.00	0.00	0.91	1.54	1.10	0.68	0.00	0.00
C10H12 Tetralin	0.87	1.33	2.05	1.11	0.92	1.13	1.03	0.93	3.48	0.49	10.08	11.88	10.02	4.77	3.08	0.97
C10H 8 Naphthalene	0.60	0.75	0.62	0.80	0.58	0.46	0.25	0.47	1.02	0.33	2.93	6.23	12.25	8.03	6.22	1.20
C10H12	0.00	0.00	0.00	0.00	0.00	0.00	0.00	0.00	0.00	0.00	0.39	1.19	0.95	0.55	0.00	0.00
C11H14 6-Methyltetralin	0.00	0.00	0.42	0.00	0.00	0.00	0.00	0.00	0.62	0.00	1.47	1.39	1.15	0.86	0.63	0.00
C11H10 2-Methylnaphthalene	0.62	1.00	0.86	0.57	0.54	0.80	0.73	0.72	0.90	0.38	1.04	2.20	2.65	2.37	2.89	1.03
C12H16 6-E ethyltetralin	0.00	0.00	0.00	0.00	0.00	0.00	0.00	0.00	0.34	0.00	0.50	0.74	0.50	0.31	0.26	0.00
C12H10 Biphenyl	0.70	1.08	1.41	0.66	0.91	0.56	0.65	0.66	2.05	0.00	3.93	1.54	0.99	0.52	0.64	0.91
C12H12 2-Ethylnaphthalene	0.59	0.95	0.72	0.77	0.87	0.00	0.31	0.27	1.28	0.48	2.49	1.46	1.44	1.47	3.16	0.82
C12H20	0.00	0.00	0.64	0.00	0.00	0.00	0.25	0.27	0.96	0.42	1.56	0.82	0.48	0.42	0.60	0.00
C12H24	0.00	0.00	0.57	0.00	0.00	0.00	0.21	0.00	0.89	0.00	1.52	0.75	0.20	0.07	0.00	0.00
C14H20 6-N-Butyltetralin	2.21	9.91	20.16	3.23	3.70	9.68	12.89	10.34	22.94	5.67	22.45	8.38	2.56	1.61	1.39	3.96
C14H15 2-N-Butylnaphthalene	1.50	4.57	5.76	2.53	2.70	4.10	3.71	3.83	6.52	3.83	10.38	7.43	3.23	3.64	4.28	3.68
C14H12	2.30	4.05	4.20	1.96	2.31	3.36	4.47	4.39	3.63	2.65	1.94	1.42	1.01	0.93	1.89	2.66
C14H12 2-Ethylbiphenyl	2.30	4.31	5.21	2.07	2.67	4.20	5.64	5.19	4.38	2.87	2.05	1.42	0.78	0.54	1.02	2.99
C14H12	0.00	0.53	0.43	0.00	0.69	0.62	0.26	0.23	0.61	0.23	0.65	1.00	0.52	0.65	0.00	0.31
C14H12 Dihydrophenanthrene	18.59	9.28	7.10	12.98	14.11	11.05	9.42	11.75	6.66	12.36	5.47	4.00	2.08	2.61	5.08	13.97
C14H12 Octahydrophenanthrene	12.79	17.96	20.53	7.78	9.56	14.10	23.29	22.85	15.04	14.19	6.35	2.55	0.99	0.35	0.21	12.11
C14H14 Tetrahydrophenanthrene	17.73	18.45	15.65	16.40	18.60	19.61	17.61	16.63	12.30	19.25	10.57	6.74	2.62	2.45	4.12	18.42
C14H10 Phenanthrene	39.86	26.02	13.78	49.81	42.49	30.50	18.91	21.17	12.39	36.48	13.89	16.32	11.85	17.63	50.87	37.82
TOTALS	101.11	104.70	105.93	101.70	101.47	102.90	104.54	104.21	122.31	103.50	112.75	180.70	299.58	364.86	178.61	102.06

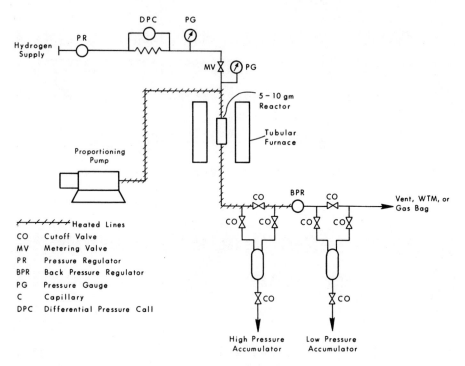

Figure 1.   Simplified flow diagram of steady-state microreactor

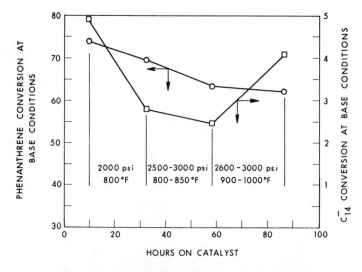

Figure 2.   Deactivation of chromia–alumina catalyst

which this may have damaged the catalyst, e.g. by reduction of
chromium oxide, is not known.

Yields of the various hydrogenation products of phenanthrene
are presented in Figure 3 and Table II. No perhydrophenanthrene
was detected in any of the products. This is consistent with the
observation that hydrogenation of the last ring of a polycon-
densed ring aromatic proceeds with considerable difficulty as
compared with the initial hydrogenation steps (18). It is
frequently supposed that the hydrogenation of polynuclear aromatics takes place ring-by-ring in a series fashion. While con-
firmation of this behavior would require more precise data at
lower conversions than are available here, it is apparent that
tetrahydrophenanthrene appears almost simultaneously with dihydrophenanthrene suggesting that the initial hydrogenations
might take place in a parallel manner.

It is of interest to compare the observed hydrogenation
product ratios with equilibrium values. Equilibrium constants
were calculated using equations available in the literature
(5,6). A hydrogen mole fraction of unity was assumed for calculation of the equilibrium product ratios, a reasonable assumption when the hydrogen treat rate greatly exceeds the rate of
hydrogen consumption. Otherwise the extent of hydrogenation at
equilibrium will be somewhat less than these calculations would
indicate. The calculated results are compared with the observed
ratios in Tables III and IV. At 800°F and 2000 psig only the
phenanthrene-dihydrophenanthrene equilibrium appears to be estab-
lished at the largest space time investigated, 1.9 $(gm/hr/gm)^{-1}$.
At the more severe conditions 1000°F and 2800 psig, the rates are
more rapid and the degree of hydrogenation is less under equilibrium conditions. All the hydrogenation products of phenanthrene, with the exception of perhydrophenanthrene, approach
their equilibrium values at large space times. The same is true
of the naphthalene-tetralin-(trans)decalin equilibria.

Inspection of Figures 4 and 5 reveals that the cracking
reactions at 800°F are limited to α-ring opening of a saturated
terminal ring. At higher temperatures dealkylation reactions
become significant. The predominant reaction at the lower
temperatures was hydrogenation to sym-octahydrophenanthrene
followed by ring opening to 6-n-butyltetralin and dealkylation to
tetralin. At the higher temperatures naphthalenes were present
in greater abundance, presumably because of the shift in thermodynamic equilibrium towards the more unsaturated species. The
yields of n-butylnaphthalene and naphthalene are comparable to
the yeilds of n-butyltetralin and tetralin respectively at 950-
1000°F. Considerable quantities of n-butylbenzene (Table II)
were observed in the products suggesting a mechanism involving
α-ring opening of tetralin. Dealkylation of the side chain was
not as clean as had been observed with the butyl substituted two
ring species. Toluene and benzene were formed in approximately
equal amounts with smaller amounts of ethylbenzene. Some xylenes

TABLE III

COMPARISON OF EQUILIBRIUM AND OBSERVED HYDROGENATION
PRODUCT RATIOS (T=800°F   P=2000 psig)

	YP4	YP5	YP6	YP2	YP3	Equili-brium
Space Time, Hr	0.30	0.57	0.96	0.96	1.95	---
$C_{14}H_{12}/C_{14}H_{10}$	0.26	0.33	0.36	0.36	0.52	0.55
$C_{14}H_{14}/C_{14}H_{10}$	0.33	0.44	0.64	0.71	1.14	2.5
$C_{14}H_{18}(sym)/C_{14}H_{10}$	0.16	0.22	0.46	0.69	1.49	6.1
$C_{14}H_{24}(L.B.P.)/C_{14}H_{8}$	0	0	0	0	0	40
$C_{10}H_{12}/C_{10}H_{8}$	1.39	1.60	2.5	1.9	3.3	12.7
$C_{10}H_{18}(cis)/C_{10}H_{8}$	0	0	0	0	0	24
$C_{10}H_{18}(trans)/C_{10}H_{8}$	0	0	0	0	0	118

TABLE IV

COMPARISON OF EQUILIBRIUM AND OBSERVED HYDROGENATION
PRODUCT RATIOS (T=1000°F   P=2800 psig)

	YP 15	YP 14	YP 13	Equilibrium
Space Time, Hr	0.16	0.29	0.98	---
$C_{14}H_{12}/C_{14}H_{10}$	0.10	0.15	0.18	0.25
$C_{14}H_{14}/C_{14}H_{10}$	0.081	0.14	0.22	0.29
$C_{14}H_{18}(sym)/C_{14}H_{10}$	0.004	0.02	0.084	0.084
$C_{14}H_{24}(L.B.P.)/C_{14}H_{10}$	0	0	0	0.018
$C_{10}H_{12}/C_{10}H_8$	0.50	0.60	0.82	1.53
$C_{10}H_{18}(cis)/C_{10}H_8$	0	0	0	0.10
$C_{10}H_{18}(trans)/C_{10}H_8$	0.15	0.19	0.30	0.37

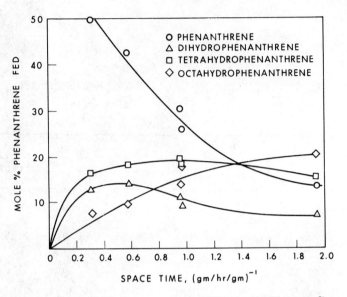

Figure 3. Hydrogenation reactions (2000 psig, 800°F nominal)

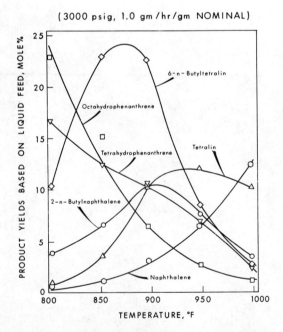

Figure 4. Effect of temperature on product distribution (3000 psig, 1.0 gm/hr/gm nominal)

were also present. Cracking within the side chain of alkyl
benzenes is indicative of a free radical cracking mechanism and
in agreement with the chromia-alumina catalyzed n-butylbenzene
dehydrocracking results reported by Csicsery (2).

As shown in Figure 5 biphenyl and 2-ethylbiphenyl were
observed in the reaction products in significant quantities.
Thus a ring opening of dihydrophenanthrene between the 10 and 11
carbons is likely taking place. This same mechanism, it is
recalled, was postulated by Penninger and Slotboom to explain
the thermal hydropyrolysis of phenanthrene. As shown in Figure
5, the yield of biphenyl at 2800 psig, 1.0 gm/hr/gm goes through
a maximum of approximately 4% at about 900°F. At these same
conditions the yields of tetralin and naphthalene were 10% and
3% (Figure 4) respectively. The decrease in biphenyl yield at
temperatures above 900°F can be explained by cracking to produce
benzene, or by the lower equilibrium concentration of dihydrophenanthrene.

The product yields by carbon number are plotted in Figure 6.
At all except the most severe conditions studied,the major component in the gas phase was n-butane. This observation is consistent with the α-ring opening and dealkylation mechanism
proposed for tetra-and octa-hydrophenanthrene cracking. At the
most severe conditions ethane was present in the greatest
quantities. This can be explained by side chain cracking of n-
butylbenzene according to the Rice-Kossiakoff mechanism or by
secondary reactions of the n-butane.

Conclusions

Three reaction paths were identified during the hydrocracking of phenanthrene over chromia-alumina. These are summarized
in Figure 7. The major reaction paths proceeded through saturation and cleavage of terminal end rings as evidence by the
presence of n-butyltetralin and n-butylbenzene intermediates and
the large quantities of n-butane in the products. However the
presence of biphenyl and 2-ethylbiphenyl in significant quantities indicates that a relatively minor reaction path involved
cracking at the saturated middle ring. Because of the importance of this latter reaction, it appears that further studies
involving nonacidic catalysts are warranted.

Acknowledgments

The financial support for this investigation was provided
by the National Science Foundation under grant GI-36597X. The
assistance of a number of individuals is gratefully acknowledged:
Dr. Stephen Billets for conducting the GC-Mass Spec analyses,
Mr. W. F. Early for aiding in the interpretation of the mass-
spec data, and Mr. K. Chandrasekhar for his assistance in the
construction and operation of the equipment. The authors would

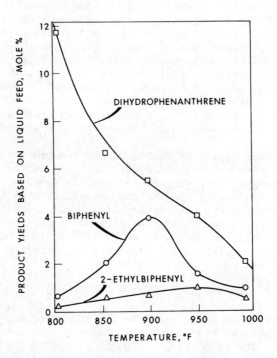

Figure 5. Effect of temperature on product distri-
bution (3000 psig, 1.0 gm/hr/gm nominal)

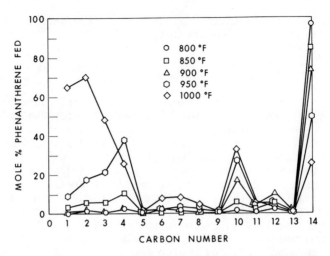

Figure 6.   Product distribution by carbon number (3000 psig, 1.0 gm/hr/gm nominal)

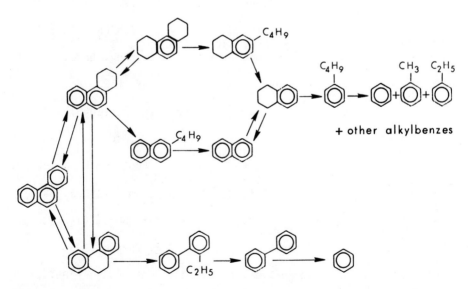

Figure 7.   Reaction paths observed during hydrocracking of phenanthrene over chromia–alumina

also like to express their appreciation to the Harshaw Chemical
Company for providing the catalyst.

ABSTRACT

This paper reports the results of an investigation of the
reactions of phenanthrene over a commerical chromia-alumina
catalyst. Chromia-alumina was selected because of its demonstrated ability to catalyze free radical reactions. Three major
reactions paths were identified. The first involved hydrogenation to sym-octahydrophenanthrene, α-ring opening of a
terminal ring and dealkylation to produce tetralin. The
tetralin reacted further by a similar pattern to produce alkyl
benzenes. The second reaction path was identical to the first
except that the initial α-ring opening occurred with tetrahydrophenanthrene. A third, relatively minor reaction path
involved saturation of the center ring, α-ring opening to produce
2-ethylbiphenyl, dealkylation and cracking at the 1,1'-position
to produce benzene.

LITERATURE CITED

(1)   Catalytic Hydrotreating of Coal Derived Liquids, Project
      Seacoke Phase II Final Report, Prepared for Office of Coal
      Research by Arco Chemical Co., Contract No. 14-01-0001-473,
      Dec., 1966.
(2)   Csicsery, S. M., J. Catalysis, (1968), 9, 416.
(3)   Csicsery, S. M., and Pines, H., J. Catalysis, (1962), 1,
      329.
(4)   Early, W. F., M.S. Thesis, University of Mississippi, 1975.
(5)   Frye, C. G., J. Chem. Eng. Data, (1962), 7, 592.
(6)   Frye, C. G., and Weitkamp, A. W., J. Chem. Eng. Data, (1969),
      14, 372.
(7)   Gardner, L. E., and Hutchinson, W. M., Ind. Eng. Chem.,
      Prod. Res. Dev., (1964), 3, 28.
(8)   Greensfelder, B. S., et al., Ind. Eng. Chem., (1945), 37,
      1168.
(9)   Griffith, R. H., Adv. Catalysis, (1948), 1, 103.
(10) Haynes, H. W., Jr., unpublished data, 1970.
(11) Penninger, M. L., and Slotboom, H. W., Erdol und Kohle,
      Erdgas, Petrochemie, (1973), 26, 445.
(12) Pines, H., and Abramovici, M., J. Org. Chem., (1969), 34,
      70.
(13) Pines H., and Csicsery, M., J. Amer. Chem. Soc., (1962),
      84, 292.
(14) Pines, H., and Goetschel, C. T., J. Org. Chem., (1965),
      30, 3530.
(15) Poole, C. P., and MacIver, D. S., Adv. Catalysis, (1967),
      17, 223.

(16) Qader, S. A., et al., ACS Div. Fuel Chem. Preprints,(1973), 18 (4), 127.

(17) Qader, S. A., et al., PREPRINTS, Div. Petrol. Chem., ACS, (1973), 18 (1), 60.

(18) Smith, H. A., "The Catalytic Hydrogenation of Aromatic Compounds", Catalysis, Vol. V, (P. H. Emmett, ed.), 175 (1957).

(19) Thomas, C. L., "Catalytic Processes and Proven Catalysts", Academic Press, New York, 1970.

(20) Turkevich, J., et al., J. Amer. Chem. Soc., (1941), 63, 1129.

(21) Voltz, S. E., and Weller, S. W., J. Amer. Chem. Soc., (1954) 76, 4701.

(22) Wiser, W. H., et al., Ind. Eng. Chem., Proc. Res. Dev., (1970), 9, 350.

(23) Wu, Wen-lung, M.S. Thesis, University of Mississippi, 1974.

5

Conversion of Complex Aromatic Structures to Alkylbenzenes

SHAIK A. QADER
Burns and Roe Industrial Services Corp., Paramus, N.J. 07652
DAVID B. McOMBER
University of Utah, Salt Lake City, Utah 84112

Alkylbenzenes (BTX) are mainly made from petroleum and coal in the United States. Petroleum
refineries produce about 95% of BTX and the rest comes
from coal. Coal based BTX is obtained as a by-product
from high temperature carbonization of coal employed
for the production of metallurgical coke. Carbonization
process yields only 2-3 gallons of BTX per ton of coal.
BTX can be produced in larger quantities by hydrogenation and hydrocracking of coal (1,2). The organic
matter of coal is mainly aromatic in nature and it
consists of clusters of each 3 to 4 benzene rings on
an average, inter-connected by aliphatic linkages.
The clusters contain mainly aromatic rings with some
hydroaromatic and naphthenic structures (3). Conversion of coal to single ring aromatics involves
release of clusters from complex organic matter with
subsequent conversion of the clusters to finished
products. Therefore, production of alkylbenzenes from
coal is due to the occurrence of hydrogenation and
hydrocracking reactions of coal and polynuclear aromatic structures released from coal. An understanding
of the hydrogenation and hydrocracking reactions of
coal, coal oil and polynuclear aromatic hydrocarbons
will, therefore, lead to the understanding of the
production of alkylbenzenes from coal. In this investigation, coal, coal oil, anthracene and phenanthrene
were hydrocracked over different catalysts in batch
systems. Product distributions, reaction kinetics and
catalyst effects were studied.

Experimental

Hydrocracking of coal, coal oil, anthracene and phenanthrene was carried out in batch stirred tank reactors (Figures 1 and 2) in the temperature range 450° - 540°c under pressures up to 3500 psi. Reactor combination shown in Figure 2 was used for the hydrocracking of coal. Hydrocracking experiments of anthracene, phenanthrene and coal oil were conducted in the same manner as described earlier (4). In case of coal, hydrogen was preheated in reactor (1) and passed into reactor (2) containing coal. No stirring was done during coal hydrocracking experiments.

A bituminous coal from Utah (Table I) was used in this work. The coal oil (Table II) used was obtained from a bituminous coal by hydrogenation using zinc chloride as the catalyst in a semi-continuous reactor system. Anthracene, phenanthrene, WS_2 and NIS used were pure grade chemicals of over 99% purity. H-zeolon was a synthetic mordenite cracking catalyst and was supplied by Norton Chemical Company. NIS-H-zeolon was prepared by spraying nickel on H-zeolon with a subsequent sulfiding operation. $NIS-WS_2-SiO_2-Al_2O_3$ catalyst used was a commercial hydrocracking catalyst. Analyses of reactants and products were done by standard methods.

Results and Discussion

Hydrocracking of anthracene and phenanthrene was done by using two different types of dual-functional catalysts, impregnated and physical mixtures. Data given in Tables III and IV indicated that the gross hydrocracking pattern of both hydrocarbons remained almost same irrespective of the nature of catalyst used. Both impregnated and physically mixed catalysts yielded almost similar liquid and gaseous products indicating that hydrocracking reaction mechanisms remained same in both cases under the experimental conditions used in this work. However, there were slight differences in actual conversions obtained with impregnated and physically mixed catalyst combinations. Product distribution data indicated that hydrocracking of anthracene and phenanthrene proceeded through the

84

HYDROCRACKING AND HYDROTREATING

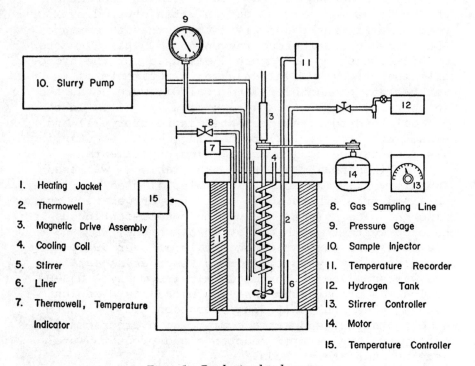

Figure 1.   Batch stirred tank reactor

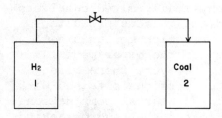

Figure 2.   Hydrocracking of coal. 1 and
2: Batch stirred tank reactor.

TABLE I
PROPERTIES OF COAL

Proximate (dry)	Wt. %
Volatile Matter	42.4
Ash	6.9
Fixed Carbon	50.7

Ultimate (dry)	
Carbon	76.42
Hydrogen	5.45
Oxygen	8.65
Sulfur	0.96
Nitrogen	1.62

TABLE II
PROPERTIES OF COAL OIL

Gravity $^\circ$API	:	12.0
Boiling Range, $^\circ$C	:	200-350
Sulfur, wt. %	:	0.82
Nitrogen, wt. %	:	0.94
Oxygen, wt. %	:	3.82
H/c (atomic)	:	0.83

86

HYDROCRACKING AND HYDROTREATING

TABLE III

Anthracene Hydrocracking
Temperature, ^{o}c : 475
Pressure, psi : 3000
Reaction Time, mts. : 20
Reactant/catalyst, wt. : 1.0

Products, %	NIS-H-Zeolon (Impregnated)	NIS-H-Zeolon (Physical Mixture)	NIS-WS2-SiO2·Al2O3 (Impregnated)	NIS-WS2-SiO2·Al2O3 (Physical Mixture)
Anthracene and Higher	54.5	55.1	51.2	52.8
Dihydro Anthracene	4.8	5.2	4.4	4.7
Tetrahydro Anthracene	15.1	15.3	16.9	16.8
Octahydro Anthracene	2.1	2.5	1.8	1.2
Hydro Anthracene Isomer (C_{14})	1.7	1.4	1.9	2.3
Hydro Anthracene Isomer (C_{13})	2.8	3.0	5.1	5.6
Naphthalene, Tetralin, Indanes	8.5	8.6	9.8	8.2
Benzene and Alkyl-benzenes	10.5	8.9	8.9	8.4
Butanes	0.8	0.3	1.9	1.7
Propane	11.5	10.5	10.7	9.6
Ethane	17.3	13.0	28.3	24.7
Methane	21.5	22.7	19.1	18.5

TABLE IV

Phenanthrene Hydrocracking
Temperature, °C : 475
Pressure, psi : 3000
Reaction time, mts. : 20
Reactant/catalyst, wt. : 1.0

Products, %	NIS-H-Zeolon (Impregnated)	NIS-H-Zeolon (Physical Mixture)	NIS-WS$_2$-SiO$_2$·Al$_2$O$_3$ (Impregnated)	NIS-WS$_2$-SiO$_2$·Al$_2$O$_3$ (Physical Mixture)
Phenanthrene and higher	22.7	21.5	24.3	25.4
Dihydro Phenanthrene	17.6	18.4	14.9	15.3
Tetrahydro Phenanthrene	26.4	25.3	27.8	28.4
Octahydro Phenanthrene	6.4	6.8	5.4	5.1
Hydro Phenanthrene Isomer (C^{14})	7.7	7.2	9.8	9.2
Hydro Phenanthrene Isomer (C^{13})	7.9	8.5	5.1	5.4
Naphthalene, Tetralin, Indanes	7.1	7.8	7.9	8.1
Benzene and Alkylbenzes	4.2	4.5	4.8	3.1
Butanes	1.2	1.5	2.8	2.3
Propane	8.9	9.5	9.2	7.1
Ethane	11.5	12.9	12.9	12.8
Methane	19.8	20.4	15.2	14.8

occurrence of hydrogenation, isomerization and hydrocracking reactions as also shown previously by Qader
et al (4-6). However, there were some differences
in the conversions and nature of products obtained
from NIS-H-zeolon and $NIS-WS_2-SiO_2-Al_2O_3$ combinations
from both anthracene and phenanthrene.

Product distribution data (Table V) obtained in
the hydrocracking of coal, coal oil, anthracene and
phenanthrene over a physically mixed NIS-H-zeolon
catalyst indicated similarities and differences
between the products of coal and coal oil on the one
hand and anthracene and phenanthrene on the other hand.
There were differences in the conversions which varied
in the order coal > anthracene > phenanthrene > coal oil.
The yield of alkylbenzenes also varied in the order
anthracene > phenanthrene > coal oil > coal under the
conditions used. The alkylbenzenes and C_1-C_4 hydrocarbon products from anthracene were similar to the
products of phenanthrene. The most predominant component of alkylbenzenes was toluene and xylenes were
produced in very small quantities. Methane was the
most and butanes the least predominant components of
the gaseous product. The products of coal and coal
oil were also found to be similar. The most predominant components of alkylbenzenes and gaseous product
were benzene and propane respectively. The data also
indicated distinct differences between products of
coal origin and pure aromatic hydrocarbons. The alkyl-
benzene products of coal and coal oil contained more
benzene and xylenes and less toluene, ethylbenzene and
higher benzenes when compared to the products from
anthracene and phenanthrene. The gaseous products of
coal and coal oil contained more propane and butanes
and less methane and ethane when compared to the
products of anthracene and phenanthrene. The differences in the hydrocracked products were obviously due
to the differences in the nature of reactants. Coal
and coal oil contain hydroaromatic, naphthenic,
heterocyclic and aliphatic structures, in addition to
polynuclear aromatic structures. Hydrocracking under
severe conditions yielded more BTX as shown in Table
VI. The yields of BTX obtained from coal, coal oil,
anthracene and phenanthrene were respectively 18.5,
25.5, 36.0, and 32.5 percent. Benzene was the most

TABLE V
NATURE OF HYDROCRACKED PRODUCTS

Catalyst: H-zeolon (10) + WS_2
Temperature, ^{o}c : 500
Pressure, psi : 3000
Reaction Time, mts. : 0 + 20
Reactant/catalyst, (wt) : 1.0

	Coal	Coal Oil	Anthracene	Phenanthrene
Conversion, wt. %	61.0	45.0	58.0	49.0
Alkylbenzenes, wt.%	2.6	3.5	8.4	4.4
Composition of Benzenes, vol. %				
Benzene	45.9	40.2	17.4	20.1
Toluene	20.2	17.5	46.0	44.2
Ethylbenzene	10.2	8.7	15.2	13.9
Xylenes	16.3	14.3	1.9	1.4
Propyl and Butylbenzenes	7.4	19.3	19.5	20.4
Composition of Gas, Vol. %				
Methane	12.4	13.2	42.6	46.4
Ethane	24.6	26.1	35.4	28.4
Propane	35.9	45.3	20.6	21.9
Butanes	27.1	15.4	1.4	3.3

TABLE VI

BTX PRODUCTION FROM DIFFERENT FEEDSTOCKS

	Coal	Coal Oil	Anthracene	Phenanthrene
Temperature, °C	540	520	520	520
Pressure, psi	3500	3000	3000	3000
Reaction Time, mts	30	30	30	30
Conversion, wt. %	82	71	79	73
BTX yield, wt. %	18.5	25.5	36.0	32.5
BTX Analysis, vol.%				
Benzene	49.5	48.4	21.5	24.8
Toluene	22.4	20.8	54.6	51.3
Ethylbenzene	9.5	10.2	19.1	18.8
Xylenes	18.6	20.6	4.8	5.1

predominant component of coal and coal oil products, whereas anthracene and phenanthrene products contained more toluene.

Kinetics

Hydrocracking conversion data were evaluated by a simple first order rate equation (i) where "x" is fractional conversion of reactant and Q is a constant. The data were found to be compatible with equation (i) as shown in Figures 3 - 6.

$$Ln(1-x) = -KT + Q \quad (i)$$

First order rate constants (Table VII) varied in the order coal > anthracene > phenanthrene > coal oil in the temperature range of 450° - 500°c under a pressure of 3000 psi. The arrhenius activation energies (Table VIII and Figure 7) based on first order rate constants indicated that the rates of hydrocracking were controlled by chemical reactions. The rate data of phenanthrene and coal oil were tested for compatibility with the dual-site mechanism of Langmuir and Hinshelwood. The model shown in equation (ii) was earlier used for testing hydrocracking data of naphthalene, anthracene and pyrene by Qader et al (5, 6).

$$\frac{1}{K^{\frac{1}{2}}} = \frac{1}{(K^* K_a K_h C_h)^{\frac{1}{2}}} + \frac{K_a^{\frac{1}{2}}}{(K^* K_h C_h)^{\frac{1}{2}}} \cdot C_r \quad (ii)$$

where K = experimental rate constant

C_h = concentration of hydrogen

C_r = concentration of reactant hydrocarbon or oil

K^*, K_a, K_h = constants

At constant hydrogen pressure, equation (ii) becomes equation (iii) since C_h becomes constant.

$$\frac{1}{K^{\frac{1}{2}}} = M + NC_r \quad (iii)$$

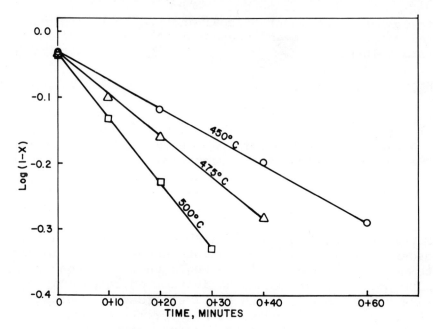

Figure 3.   First order plot of anthracene

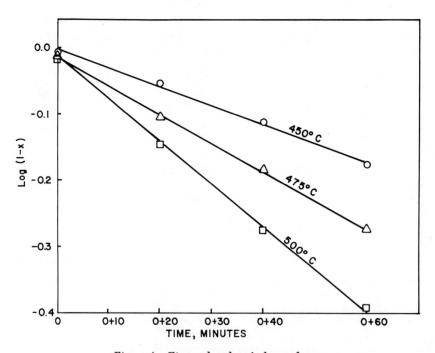

Figure 4.   First order plot of phenanthrene

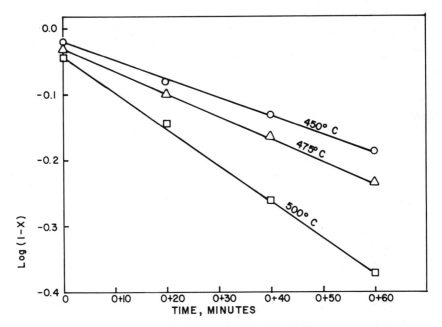

Figure 5.   First order plot of coal oil

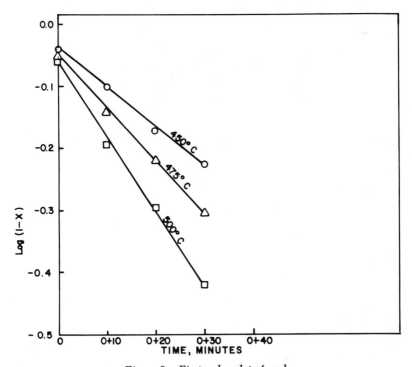

Figure 6.   First order plot of coal

TABLE VII

FIRST ORDER RATE CONSTANTS $(Mint.^{-1})$
Pressure : 3000 psi (Hot)

Temperature, °C	Coal	Anthracene	Phenanthrene	Coal Oil
450	0.0142	0.0098	0.0065	0.0063
475	0.0196	0.0149	0.0101	0.0089
500	0.0276	0.0230	0.0147	0.0127

TABLE VIII

ARRHENIUS ACTIVATION ENERGIES

	Activation Energy (K cal/gm moles)
Coal	16.58
Coal Oil	17.44
Anthracene	19.30
Phenanthrene	20.68

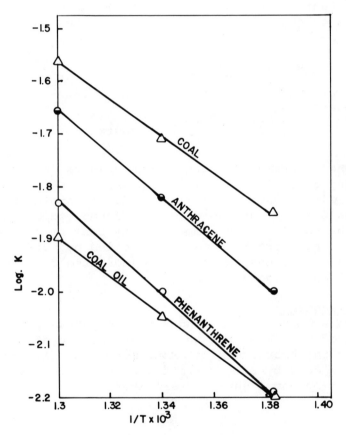

Figure 7.   Arrhenius plots

Where $M = \dfrac{1}{(K^* \, K_a K_h C_h)}^{\frac{1}{2}} = $ constant

$N = \left(\dfrac{K_a}{K^* \, K_h C_h}\right)^{\frac{1}{2}} = $ constant

At constant concentration of hydrocarbon or oil, C_r becomes constant and equation (ii) becomes equation (iv)

$$K = Z \cdot C_h$$

Where $Z = \dfrac{K^* \, K_a K_h}{(1 + K_a \, C_r)^2} = $ constant (iv)

The experimental data obtained in the hydro-cracking of coal, coal oil and phenanthrene were tested with equations (iii) and (iv) as shown in Figures 8 and 9 and the data were found compatible with the model. This compatibility confirmed that the rates of hydrocracking were controlled by surface chemical reactions as were also indicated by arrhenius activation energies (Table VIII).

Acknowledgement

This work was done under the sponsorship of the Office of Coal Research and the University of Utah. Some of the hydrocracking experiments on coal and coal oil were done by Bill Berrett and the analytical work was done by Jim Light and Earl Everett.

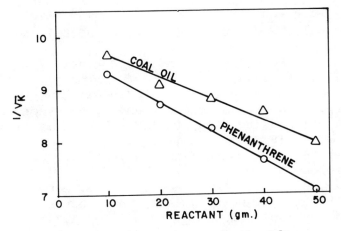

Figure 8.   *Variation of rate constant with weight*

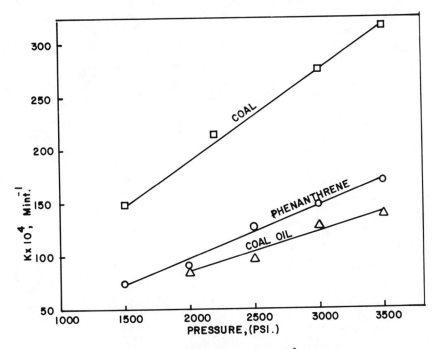

Figure 9.   *Variation of rate constant with pressure*

Literature Cited

1. Schroeder, W.C., U. S. Patent 30,030,297 and 3,152,063
2. Qader, S. A. - The Oil and Gas Journal, 72, No. 29, 58 (1974)
3. Given, P. H. - Fuel, 39, 147 (1960)
4. Qader, S. A., D. B. McOmber and W. H. Wiser, 166th National Meeting, ACS, Fuel Chemistry Division, preprints 18, No. 4, 127 (1973)
5. Qader, S. A. - Journal of the Institute of Petroleum 59, No. 568, 178 (1973)
6. Qader, S. A., L. Chun Chen and D. B. McOmber, 165th National Meeting, ACS, Fuel Chemistry Division - preprints 18, No. 1, 60 (1973)

Diesel and Burner Fuels from Hydrocracking in *Situ Shale* Oil

PHILIP L. COTTINGHAM and LEO G. NICKERSON

Laramie Energy Research Center, Energy Research and Development Administration, Laramie, Wyo. 82070

For several years, the Laramie Energy Research Center has conducted research in recovering shale oil through in situ retorting by the underground combustion method. Crude shale oils produced by this method normally have lower specific gravities, viscosities, and pour points than do crude shale oils produced in the gas combustion or N-T-U retorts. They also contain somewhat less nitrogen, but the sulfur content is not greatly different from that of the crude oils from these aboveground retorts. Percentages of both elements are higher than desirable in liquid fuels or in feedstocks that are to be processed to high-quality liquid fuels by most catalytic refining processes.

Catalytic hydrogenation at cracking conditions, referred to here as "hydrocracking," has been found by previous experiments to be a suitable method for producing low-sulfur, low-nitrogen "synthetic crude" oils from in situ crude shale oils ([1,2,3]). One of the primary objectives of these previous experiments has been to increase the quantity of gasoline contained in the hydrocracked synthetic crudes. A limited investigation of the suitability of fractions of the hydrocracked oils as catalytic reforming and catalytic cracking feedstocks for the production of gasoline has been made ([1,2]); however, little attention has been given to the types of diesel fuels and fuel oils that can be prepared from the fractions.

The purpose of the present work was to investigate the quantity and quality of diesel fuels and fuel oils that could be prepared from the liquid product obtained by hydrocracking of in situ combustion shale oil. A sample of the in situ crude oil was hydrocracked at operating conditions that had been found by previous experiments to be effective in eliminating most of the nitrogen and sulfur from the oil and in greatly reducing the average boiling point of the oil. The hydrocracked oil was distilled into a 350° F end-point reforming stock and small-volume heavier fractions. Diesel fuels, suitable for use also as fuel oils, were prepared by blending small-volume fractions.

Apparatus and Operating Procedure

A simplified flow diagram of the hydrocracking equipment is shown in Figure 1. The reactor was a type 316 stainless steel tube 40 inches long, with 9/16-inch inside diameter and 1-inch outside diameter. A stainless steel screen 10½ inches from the bottom of the reactor supported 50 ml of catalyst that extended approximately 12¼ inches above the screen. A second screen on top of the catalyst supported 50 ml of quartz chips in the upper, pre-heating section of the reactor. The reactor was surrounded by a cylindrical, 3-inch-outside-diameter aluminum block, contained in a 36-inch-long electric furnace that covered essentially all but the exposed high-pressure fittings at the ends of the reactor. Temperatures were measured by five thermocouples placed at uni-formly spaced intervals along the catalyst-containing section of the reactor in a groove in the inner wall of the aluminum block. Other thermocouples were used to control the temperature of the surrounding furnace.

Before the hydrocracking experiment, the catalyst was pre-treated in the reactor for 16 hours at 700° F with a mixture of 1.1 standard cubic feet of air and 0.41 pound of steam per pound of catalyst per hour. After the pretreatment, the reactor was cooled to 500° F; then the steam was cut off. When the reactor cooled to 350° F, the air flow was cut off. The reactor was next purged with helium and pressurized to 250 psig with hydrogen. A stream of hydrogen containing 5 volume-percent hydrogen sulfide at a total pressure of 250 psig was then passed through the reactor at a rate of 1 scf per hour for 4 hours. During this time the reactor was heated from 350° to 600° F. The reactor was then heated to the planned hydrocracking temperature, and brought to the planned total reaction pressure with hydrogen, and a stream of hydrogen at the proposed experimental rate was passed through the reactor for 30 minutes. The oil feed pump was then started.

During the experiment, oil pumped by positive displacement was mixed with hydrogen at the top of the reactor and passed down-ward through the catalyst. The resulting products passed through a backpressure regulator into a separator maintained at 200 psig and 75° F. Gas from the separator passed through a second back-pressure regulator and was metered and sampled for gas chromato-graphic analysis. The liquid product was drained from the separa-tor at the end of each 24-hour period and was washed with water to remove hydrogen sulfide and ammonia before it was analyzed. Hy-drogen flow was metered with a mass flowmeter, but total hydrogen feed was measured by the volume removed from calibrated storage vessels. Hydrogen consumption was calculated from the analyses of feed and product gases.

After the experiment, the used catalyst was purged with steam and then regenerated with a mixture of steam and air; the regener-ation gas was dried and any carbon dioxide formed was absorbed by Ascarite for determination of the percent carbon deposit.

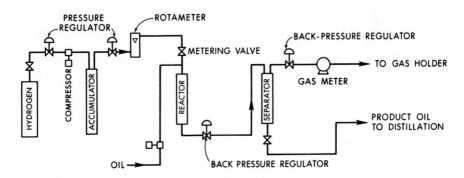

Figure 1.   Simplified flow diagram of hydrogenation unit

TABLE I. - Hydrocracking of in situ crude shale oil
over nickel-molybdena catalyst

Operating conditions:		
Temperature, ^{0} F	800	
Pressure, psig	1500	
Space velocity, V_O/V_C/hr	0.48	
Throughput, V_O/V_C	125	
Hydrogen feed, scf/bbl	5000	
Hydrogen consumed, scf/bbl	1760	

	In situ crude	Hydrocracked oil
Yield, volume-percent	100.0	97.6
Yield, weight-percent	100.0	89.7
Properties:		
Specific gravity, 60/60^0 F	0.8800	0.8086
^{0} API	28.75	43.56
Nitrogen, weight-percent	1.62	0.02
Sulfur, weight-percent	0.82	0.02
Carbon/hydrogen weight ratio	7.01	ND
Carbon residue, weight-percent	1.8	ND
Pour point, ^{0} F	40	ND
Viscosity at 100^0 F:		
Kinematic, cs	5.82	ND
Saybolt Universal Seconds	45	ND
ASTM distillation, ^{0} F:		
IBP	293	138
5 volume-percent recovered	346	258
10 volume-percent recovered	373	305
20 volume-percent recovered	424	362
30 volume-percent recovered	461	403
40 volume-percent recovered	491	437
50 volume-percent recovered	536	471
60 volume-percent recovered	584	507
70 volume-percent recovered	621	546
80 volume-percent recovered	658	599
90 volume-percent recovered	-	672
EP	704	717
Recovery, volume-percent	92.0	94.5
Residue, volume-percent	6.0	2.5
Loss, volume-percent	2.0	3.0

ND, not determined

Preparation of Hydrocracked Oil

The in situ crude shale oil feedstock was obtained during the 28th through the 37th day of operation of a 42-day in situ combustion retorting experiment at a site near Rock Springs, Wyo. (4,5,6). The crude contained 0.01 weight-percent ash, which was removed by filtration with diatomaceous silica filter aid before the oil was hydrocracked. The filtered oil was hydrocracked in a once-through operation at 800° F, 1,500 psig, and 0.48 volume of oil per volume of catalyst per hour (V_o/V_c/hr) over a catalyst that contained nickel oxide and molybdenum oxide on alumina in the form of 1/8-inch by 1/16-inch extrusions. Total throughput of crude oil in the planned 10-day run was 125 volumes of oil per volume of catalyst (V_o/V_c). Operating conditions, properties of the feedstock, yields, and properties of the hydrocracked oil (total liquid product) are shown in Table I.

The yield of synthetic crude was 97.6 volume-percent, or 89.7 weight-percent. The decrease in quantity of liquid was accompanied by considerable reduction in the average boiling point, and specific gravity of the liquid product compared with those of the in situ crude. At the same time, about 98 percent of the nitrogen and sulfur content was eliminated. Gas formed was 9.4 weight-percent of the feed, and carbon deposited on the catalyst was 0.13 weight-percent of the feed.

The analysis of the product gas, calculated to a hydrogen-free basis, is shown in Table II. The high percentage of methane and the predominance of n-butane and n-pentane over the branched compounds suggest that they were produced by a thermal cracking reaction with hydrogenation of the olefins.

TABLE II. - Analysis of gas from hydrocracking in situ crude

Gas	Volume-percent of gas
Methane	45.8
Ethane	33.6
Propane	14.0
i-Butane	1.2
n-Butane	4.2
i-Pentane	0.5
n-Pentane	0.7

Diesel Fuels and Fuel Oil

As a means of preparing high percentages of diesel fuels with good quality from the hydrocracked oil, a batch distillation of the oil was made in a packed laboratory distillation column in which a first fraction was removed at a corrected column head

temperature of 364° F (at 760 mm). This fraction, considered as a
catalytic reforming feedstock, had an ASTM-distillation (7) end
point of 350° F. Its Research Method octane numbers were only 39,
clear, and 69 with 3 ml of tetraethyllead, and it would require
reforming before use in gasoline. The remainder of the distillation was conducted at 10 mm absolute pressure, and 54 small-volume
distillate fractions were collected at 5° F intervals between distillation column head temperatures of 145° F and 428° F (at 10 mm),
corresponding to temperatures of 364° and 725° F at 760 mm.

    The distillation residuum contained a large amount of wax
that separated when the residuum was cooled to room temperature.
The wax was removed by mixing the residuum with three times its
volume of acetone, chilling the mixture to 32° F, and filtering
out the precipitated wax. Acetone was removed from both the fil-
trate and precipitate by distillation. Total wax removed, including a small quantity washed from the distillation column with
boiling acetone, amounted to 48 weight-percent of the residuum.

    From a study of the distillation data for the hydrocracked
oil, it was estimated that quantities of city bus (C-B), truck-
tractor (T-T), and stationary-marine (S-M) diesel fuels in the
volume ratios of 2:2:1 and fitting the average distillation characteristics (8) of diesel fuels now being sold could be made from
the hydrocracked small-volume distillate fractions. Beginning
with the type C-B diesel fuel, appropriate amounts of the small-
volume fractions were calculated that would produce a fuel approximately fitting the 10, 50, and 90 percent ASTM distillation
points of typical diesel fuels now being sold. Similar calculations were made for the desired types T-T and S-M fuels. The calculated quantity of each fuel was then adjusted to use up as much
as possible of the available small-volume fractions without greatly changing the desired distillation characteristics of the proposed fuels. According to these calculations, portions of some of
the small-volume fractions would be left over, and these leftover
portions were considered suitable for mixing with the dewaxed distillation residuum to make fuel oil.

    The small-volume fractions were blended according to the
above procedure, producing the quantities of diesel fuels shown
(as percentages) in Table III. The dewaxed residuum was mixed
with approximately four times its volume of lower boiling oil not
used in the diesel fuel blends, and is reported in Table III as
fuel oil. Properties of the blended diesel fuels and fuel oil are
shown in Table IV.

    The total yield of diesel fuels was 51.6 volume-percent of
the in situ crude. The properties of these fuels fell within the
limits (Table V) of those of corresponding petroleum diesel fuels
currently marketed in the United States (8) except for the carbon
residue of the S-M shale-oil diesel fuel; this residue was slightly higher than those of the petroleum diesel fuels but was probably acceptable. The value of 0.36 weight-percent for the carbon
residue on the 10-percent bottoms of the S-M fuel was only a

TABLE III. - Yields of products from distillation and blending

	Weight-percent of in situ crude	Volume-percent of in situ crude
Reforming stock	16.0	18.5
C-B diesel fuel	19.7	21.3
T-T diesel fuel	18.5	19.6
S-M diesel fuel	10.1	10.7
Fuel oil	20.4	21.5
Wax	3.8	ND
Light ends from distillation	0.6	ND
Distillation loss	0.6	ND

ND, not determined

trifle greater than the upper ASTM limit of 0.35 percent for the lighter grade 2-D diesel fuels (7), and no limit is listed for the grade 4-D which corresponds to type S-M. With the exception of the carbon residue and high flash point, the properties of the S-M shale-oil diesel fuel also fell within the limits of the properties of the type R-R (railroad) diesel fuels now being marketed. Most of the properties of the fuel oil blend containing the de-waxed residuum were within the range of properties of the petroleum S-M diesel fuels shown in Table V. Its high cetane index of 65 suggests that this fuel oil be used as an S-M diesel fuel rather than as a burner fuel.

The sulfur percentages of 0.02 weight-percent or less in the shale oil diesel fuels would permit these fuels to be classed as low-sulfur diesel fuels. Nitrogen percentages were low, ranging from 141 to 202 parts per million (ppm). Cetane indexes, calculated by ASTM method D-976 (7), were excellent, ranging from 48 to 56.

Table VI shows the range of properties of grades 1, 2, and 4 burner fuel oils marketed in the United States in 1974 (9). Most of these fuels were also marketed as diesel fuels (8,9). The shale-oil types C-B and T-T diesel fuels would, respectively, fit with the grades 1 and 2 burner fuels. The shale-oil S-M diesel fuel would also fit with the grade 2 burner fuels. The shale-oil fuel oil fraction had a distillation range resembling the grade 2 fuels, but its viscosity was intermediate between those of the grade 2 and grade 4 fuels.

The C-B, T-T, and S-M shale oil diesel fuels respectively met the ASTM requirements (7) for grades 1-D, 2-D, and 4-D diesel fuels and Nos. 1, 2, and 4 fuel oils, with some minor exceptions. Thus, the 90-percent distillation temperature of the T-T diesel fuel was 3° F lower than the ASTM lower limit for grade 2-D diesel and No. 2 fuel oil. Also, the viscosity of the S-M diesel fuel was slightly lower than the lower limit specified for grade 4-D

TABLE IV. - Properties of reforming stock, diesel fuels,
           and fuel oil

	Reforming stock	C-B diesel	T-T diesel	S-M diesel	Fuel oil
Specific gravity, 60/60° F	0.7600	0.8159	0.8262	0.8300	0.8367
° API	54.68	41.93	39.75	38.98	37.62
Flash point, ° F	NAp	188	212	265	325
Pour point, ° F	NAp	-50	-30	-20	30
C residue on 10 pct bottoms, wt-pct	NAp	0.14	0.16	0.36	0.42
Ash, wt-pct	NAp	<0.001	<0.001	<0.001	<0.001
Viscosity at 100° F:					
Kinematic, cs	NAp	1.66	2.40	2.96	5.07
SUS	NAp	32	34.0	35.8	42.6
Sulfur, wt-pct	<0.001	0.01	0.02	0.02	0.02
Nitrogen, ppm	20	141	166	202	396
Cetane index	NAp	48	54	56	65
ASTM distillation, ° F at 760 mm:					
IBP	158	379	404	410	394
10 vol-pct recovered	223	400	442	450	528
20 vol-pct recovered	246	409	454	468	562
30 vol-pct recovered	263	414	465	483	572
40 vol-pct recovered	277	418	474	495	580
50 vol-pct recovered	290	422	484	507	588
60 vol-pct recovered	302	429	494	524	599
70 vol-pct recovered	313	437	505	546	610
80 vol-pct recovered	323	447	518	573	624
90 vol-pct recovered	335	469	537	605	666
EP	350	517	554	628	667
Recovery, vol-pct	97.0	95.0	98.0	97.5	92.0
Residue, vol-pct	2.0	2.0	1.0	2.0	7.0
Loss, vol-pct	1.0	3.0	1.0	0.5	1.0

NAp, not applicable

TABLE V. - Range of diesel fuels produced
in the United States during 1974

Test	Type C-B Min.	Type C-B Max.	Type T-T Min.	Type T-T Max.	Type R-R Min.	Type R-R Max.	Type S-M Min.	Type S-M Max.
Gravity, °API	33.3	47.4	27.3	47.4	27.2	43.4	9.6	41.1
Flash point, °F	124	194	112	230	136	198	136	290
Viscosity at 100° F:								
Kinematic, cs	1.50	4.34	1.30	4.55	1.60	4.34	1.94	22.1
Pour point, °F	-70	25	<-60	25	-40	25	-35	65
Sulfur, weight-percent	0.004	0.52	0.001	1.07	0.01	0.85	0.06	2.85
C res. on 10 percent, weight-percent	0.000	0.18	0.001	0.26	0.014	0.26	0.05	0.18
Ash, weight-percent	0.000	0.005	0.000	0.010	0.000	0.005	0.000	0.02
Cetane no. or index	39.8	65.8	38.1	65.8	34.1	65.8	38.0	63.1
Distillation temperature, °F:								
IBP	320	488	301	488	310	488	327	535
10 percent recovered	357	520	329	520	376	520	384	595
50 percent recovered	403	558	382	596	434	558	454	680
90 percent recovered	454	603	444	656	476	639	535	656
EP	476	663	476	700	514	688	582	700

TABLE VI. – Range of properties of distillate burner fuels produced in the United States during 1974

Test	Grade 1		Grade 2		Grade 4	
	Min.	Max.	Min.	Max.	Min.	Max.
Gravity, ° API	36.2	47.9	29.5	43.0	17.6	28.2
Flash point, ° F	104	160	114	192	156	216
Viscosity at 100° F:						
Kinematic, cs	1.28	2.60	1.70	3.70	9.64	30.1
Pour point, ° F	-70	0	-50	25	-50	75
Sulfur, weight-percent	0.002	0.38	0.03	0.64	0.46	1.44
C res. on 10 percent, weight-percent	0.00	0.18	0.002	0.27	NR	NR
C res. on 100 percent, weight-percent	NR	NR	NR	NR	0.52	7.6
Ash, weight-percent	NR	NR	NR	NR	0.002	0.03
Water and sediment, volume-percent	0.000	0.025	0.00	0.10	0.06	0.2
Distillation temperature, ° F:						
IBP	303	380	293	416	NR	NR
10 percent recovered	336	413	333	469		
50 percent recovered	382	480	432	544		
90 percent recovered	440	560	528	644		
EP	464	614	575	698		

diesel fuel and No. 4 fuel oil. The pour points met the ASTM requirements for burner fuel oils. The ASTM pour point requirements for diesel fuels depend on the ambient temperature at the place of use; for most purposes, the pour points of the shale-oil diesel fuels would be satisfactory.

The properties of the fuel oil blend met the requirements for ASTM No. 4 fuel oil (usually considered a distillate fuel) except that its viscosity was low, falling between the specified requirements for No. 2 and No. 4 fuel oils (7), and its pour point of 30° F was higher than the ASTM-recommended maximum of 20° F. The fuel could be used as a low-sulfur, high-cetane-index grade 4-D diesel fuel in warm weather or where preheating facilities were available.

The yield of fuel oil prepared from the blended dewaxed residuum was 21.5 volume-percent of the in situ crude. The total of this, plus the 51.6 volume-percent of diesel fuels, amounted to 73.1 volume-percent of the in situ crude that could be used as low-sulfur diesel fuels or Nos. 1 through 4 burner fuels.

Summary

In situ crude shale oil, produced by the underground combustion retorting method, was hydrocracked over a nickel-molybdena catalyst at 800° F and 1,500 psig. The liquid product was distilled into a low-octane reforming feedstock amounting to 18.5 volume-percent of the in situ crude plus a large number of small-volume higher boiling fractions and a small quantity of waxy residuum. Most of the small-volume distillate fractions were blended into types C-B, T-T, and S-M diesel fuels whose properties resembled those of diesel fuels and distillate fuels now being marketed. The diesel fuels also met the ASTM requirements for diesel fuels and distillate fuel oils with minor exceptions. All of the shale-oil diesel fuels had very low sulfur contents and very high cetane indexes.

The dewaxed distillation residuum was blended with a portion of the small-volume distillate fractions to form a fuel oil fitting the requirements for No. 4 fuel oil except that its viscosity was between the viscosity limits for No. 2 and No. 4 fuels and its pour point of 30° F was higher than ASTM recommendations, although in the range of No. 4 fuel oils now being marketed. This fuel oil could also be used as a low-sulfur, high-cetane-index S-M diesel fuel.

The yield of diesel fuels was 51.6 volume-percent of the in situ crude, and the yield of No. 4 fuel oil was 21.5 volume-percent, for a total of 73.1 volume-percent of the in situ crude that could be used as low-sulfur, high-cetane-index diesel fuels or Nos. 1 through 4 burner fuels.

110

HYDROCRACKING AND HYDROTREATING

Literature Cited

1. Frost, C. M., and P. L. Cottingham. BuMines RI 7844 (1974).
2. Frost, C. M., R. E. Poulson, and H. B. Jensen, Preprints, Div. of Fuel Chem., Inc., ACS (1974) 19 (2) 156.
3. Poulson, R. E., C. M. Frost, and H. B. Jensen, Preprints, Div. of Fuel Chem., Inc., ACS (1974) 19 (2) 175.
4. Burwell, E. L., H. C. Carpenter, and H. W. Sohns. BuMines TPR 16 (1969).
5. Burwell, E. L., T. E. Sterner, and H. C. Carpenter. J. Petrol. Technol. (1970) 1520.
6. Carpenter, H. C., E. L. Burwell, and H. W. Sohns. J. Petrol. Technol. (1972) 21.
7. American Society For Testing Materials. 1973 Book of ASTM Standards. Part 17: Petroleum Products.
8. Shelton, Ella Mae. BuMines PPS 87 (1974).
9. _____. BuMines PPS 86 (1974).

Synthoil Process and Product Analysis

HEINZ W. STERNBERG, RAPHAEL RAYMOND, and SAYEED AKHTAR

Energy Research and Development Administration, Pittsburgh Energy Research Center, 4800 Forbes Ave., Pittsburgh, Penn. 15213

Many power generating stations in the United States burn imported fuel oils, instead of coal, to meet environmental protection regulations. Large quantities of fuel oils are imported annually at considerable strain to the nation's balance of payment, a strain that could be removed by burning environmentally acceptable coal-derived fuels prepared from internal resources. Towards this end, the Energy Research and Development Administration is developing a process known as the SYNTHOIL process to convert coal to a low-sulfur, low-ash utility fuel. In the process, coal is liquefied and hydrodesulfurized catalytically in a turbulent-flow, packed-bed reactor (1-3). The gross liquid product on centrifugation yields a nonpolluting fuel oil suitable for power generation. The results of a study of the chemical composition and viscosity of the product oils obtained at different operating conditions are presented in this paper. The interrelationship of chemical composition and viscosity is of special interest in view of the importance of the latter in centrifugation of the gross liquid products and in pipeline transportation of the product oil.

Plant and Operating Conditions

The flowsheet of a 1/2-ton (slurry) per day SYNTHOIL bench scale plant, currently in operation at the Energy Research and Development Administration laboratory in Bruceton, Pennsylvania, is shown in figure 1. The vertically placed reactor is made of two interconnected stainless steel tubings of 1.1-inch ID x 14.5-ft long each. The upper end of the first section is connected to the lower end of the second section with a 5/16-inch ID empty tubing. Thus, the plant may be operated with one or both sections of the reactor packed with catalyst while the fluids flow upwards through each.

Hydrogen and a slurry of 35 to 45 percent coal in recycle oil are introduced concurrently in a 3-inch ID x 11-ft long preheater packed with 3/4-inch x 3/4-inch ceramic pellets which

promote heat transfer. The heated feed stream enters the reactor
at the lower end of the first section and the product stream
exits at the upper end of the second section. The liquids and
unreacted solids are separated from the gases and centrifuged
to obtain the product oil. The gases, after scrubbing out H_2S
and NH_3, are recycled without depressurization. The flow of
H_2 through both the preheater and the reactor is in turbulent
regime.

The operating conditions for the three runs FB-38, FB-39,
and FB-40, the results from which are discussed in this paper,
are given in table I. The length of the 1.1-inch ID reactor
was 29 ft in run FB-38 and 14.5 ft in FB-39 and FB-40. Runs
FB-38 and FB-39 were both conducted at 4,000 psi but the reactor
temperature was 415° C in FB-38 and 450° C in FB-39. Run FB-40
was conducted at 2,000 psi and 450° C.

TABLE I.- Run numbers and operating conditions

Run No.	FB-38	FB-39	FB-40
Length of the 1.1-inch ID reactor, ft	29	14.5	14.5
Charge weight of Co-Mo/SiO$_2$-Al$_2$O$_3$ catalyst, lb	11	5.5	5.5
Plant pressure, psig	4,000	4,000	2,000
Reactor temperature, ° C	415	450	450
Feed rate of (35 coal + 65 recycle oil) slurry, lb/hr	25	25	25
Time on stream, hr	30	380	360

The analysis of the coal used in these runs is given in
table II. The coal contained 5.5 percent sulfur and 16.5 percent
ash. The pyritic sulfur in coal was 3.08 percent and the organic
sulfur 1.95 percent.

Product Analysis

The liquid products were collected in 4-hour batches, centrifuged, sampled, and analyzed for asphaltene, oil, elementary
composition (C, H, N, S), ash, specific gravity, and viscosity.
In runs FB-38 and FB-40, the asphaltene content increased
with time. In run FB-39, on the other hand, the asphaltene
content decreased during the first 120 hours from 18 percent
to 7 percent, remained around 7 percent for about 60 hours, and

TABLE II.- <u>Analysis of feed coal,[1] as received</u>

Proximate analysis, wt pct
  Moisture 4.2
  Ash. 16.5
  Volatile matter. 36.2
  Fixed carbon 43.1

Ultimate analysis, wt pct
  Hydrogen 4.8
  Carbon 60.7
  Nitrogen 1.2
  Oxygen 11.3
  Sulfur 5.5
  Ash. 16.5

Forms of sulfur, wt pct
  Sulfate. 0.47
  Pyritic. 3.08
  Organic. 1.95

Calorific value, Btu/lb 11,020

Rank: hvBb

[1] A blend from Kentucky seams #9, 11, 12, and 13 which are mined together. Ohio County, Western Kentucky.

114   HYDROCRACKING AND HYDROTREATING

during the last 200 hours increased from 7 percent to 25 percent.
   The analyses of some of the samples collected from runs
FB-38, FB-39, and FB-40 are listed in tables III, IV, and V.
Inspection of the data shows that the viscosity of the product
oil increases with asphaltene content. To learn more about the
nature of asphaltenes and their effect on viscosity, we fraction-
ated samples of product oil from runs FB-38 and FB-40 according
to the scheme shown in figure 2. Ultimate analyses and molecular
weights of the various fractions are shown in tables VI, VII,
and VIII. In addition to heavy oil and asphaltene, these tables
also list pyridine sols and pyridine insols. These two fractions
are always lumped together as "benzene insols", but we separated
the benzene insols (toluene insols in our case) into a pyridine
soluble and a pyridine insoluble fraction. It is a common practice to measure the efficiency of a coal liquefaction process
by the percent coal-to-benzene sols conversion on a moisture-
and ash-free basis, the assumption being that the benzene insoluble material is unreacted coal and, by implication, that
it is not soluble in the product oil and does not contribute
to its viscosity. These assumptions are misleading. Tables
VI, VII, and VIII show that in all cases more than 50 percent
of the toluene insols are pyridine soluble and therefore cannot
be considered unreacted coal. It is only the pyridine insols
that may be unreacted coal, carbon, or ash. Moreover, removal
of the toluene insols results in a considerable decrease in the
viscosity of the product oil. For example, the viscosity of
a product oil (FB-40, batch 88) dropped from 1,433 to 907 after
removal of 8.1 percent toluene insols. It is reasonable to assume
that of the 8.1 percent toluene insols removed, only the 4.8
percent pyridine sols are responsible for this decrease in
viscosity. A further and much larger decrease in viscosity
occurred when the asphaltenes, representing 40.2 percent of the
sample, were removed. The viscosity then dropped from 907 to
16, as shown in figure 3.

TABLE III.-   Analyses of centrifuged liquid product samples
                      taken during run FB-38

Batch No.	Sp gr, 60°F/ 60°F	Vis- cosity, SSF at 180° F	OBI[1]	As- phal- tenes	Oil	Ash	C	H	N	S
1	1.061	39.4	2.6	21.2	75.3	0.9	87.4	9.2	0.8	0.36
4	1.064	44.3	2.2	22.8	73.8	1.2	86.6	9.1	0.8	0.42
7	1.079	85.6	2.8	24.3	70.9	2.0	85.9	8.9	0.8	0.56

Analyses, wt pct

[1] Organic benzene insolubles

TABLE IV.- Analyses of centrifuged liquid product samples taken from run FB-39

Batch No.	Sp gr, 60°F/60°F	Viscosity SSF	OBI[1]	Asphaltenes	Oil	Ash	C	H	N	S
					Analyses, wt pct					
1	1.082	21.5 at 180° F	2.4	18.0	79.1	0.50	88.8	8.5	0.8	0.30
31	1.023	29.1 at 77° F	0.4	7.2	92.3	0.01	88.7	9.5	0.6	0.17
49	1.034	67.2 at 77° F	0.9	11.4	87.6	0.10	88.1	9.3	0.8	0.14
91	1.094	32.5 at 180° F	1.6	25.4	71.9	1.06	86.5	8.4	1.2	0.51

[1] Organic benzene insolubles

TABLE V.- Analyses of centrifuged liquid product samples taken during run FB-40

Batch No.	Sp gr, 60°F/60°F	Viscosity, SSF at 180° F	OBI[1]	Asphaltenes	Oil	Ash	C	H	N	S
					Analyses, wt pct					
1	1.060	13.5	1.1	17.3	81.4	0.2	88.5	8.6	1.0	0.22
43	1.130	56.2	4.6	29.4	64.3	1.7	86.5	7.5	1.5	0.55
90	1.146	75.7	5.7	28.9	62.5	2.9	85.3	7.3	1.6	0.71

[1] Organic benzene insolubles

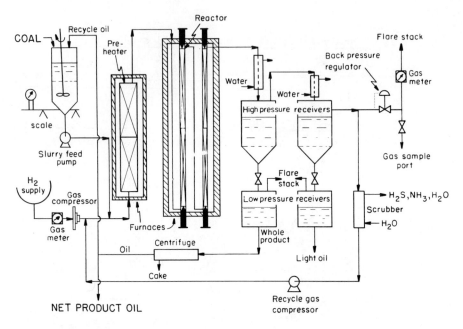

Figure 1.   Synthoil pilot plant flow sheet

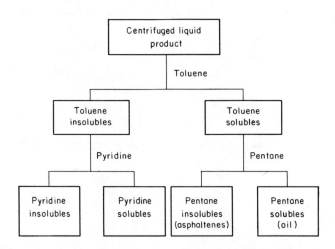

Figure 2.   Fractionation scheme for centrifuged liquid product
(CLP)

TABLE VI.— Analysis of centrifuged liquid product (CLP) from run FB-38, batch 7 (415° C, 4,000 psi). Viscosity of CLP: 85.6 SSF at 180° F

Fraction	Pct of Total product	Ultimate analysis, wt pct							Mol wt	Viscosity[1]
		C	H	N	S	Ash	O	H/C		
Oil	71.7	89.01	9.69	0.43	0.18	0.01	0.68	1.30	289	53 at 100° F
Asphaltenes	23.5	85.44	6.78	1.12	0.77	0.36	5.53	0.95	671	
Pyridine solubles	2.6	81.97	5.69	2.03	0.91	2.68	6.72	0.83		
Pyridine insolubles	2.2					71.9				

[1] In centistokes

TABLE VII.— Analysis of centrifuged liquid product (CLP) from run FB-40, batch 7 (450° C, 2,000 psi).

Viscosity of CLP: 58.9 centistokes at 140° F, 14.3 SSF at 180° F

| Fraction | Pct of total product | Ultimate analysis, wt pct | | | | | | Mol wt | Viscos-ity[1] |
		C	H	N	S	Ash	O	H/C		
Oil	75.2	88.52	8.89	0.83	0.23	0.0	1.53	1.20	227	11.8 at 140° F
Asphaltenes	22.3	89.03	6.13	2.10	0.27	0.0	2.47	0.82	459	
Pyridine solubles	1.8	86.80	4.75	2.63	0.46	2.29	3.07			
Pyridine insolubles	0.7					28.5				

[1] In centistokes

TABLE VIII.— Analysis of centrifuged liquid product (CLP) from run FB-40, batch 88 (450° C, 2,000 psi).
Viscosity of CLP: 1,433 centistokes at 140° F, 97.6 SSF at 180° F

Fraction	Pct of total product	Ultimate analysis, wt pct						H/C	Mol wt	Viscosity[1]
		C	H	N	S	Ash	O			
Oil	51.7	87.57	8.49	1.10	0.42	0.0	2.42	1.16	225	16.5 at 140° F
Asphaltenes	40.2	87.33	6.27	2.14	0.62	0.0	3.64	0.86	417	
Pyridine solubles	4.8	73.06	4.39	2.32	1.41	14.28	4.54	0.72		
Pyridine insolubles	3.3					65.5				

[1] In centistokes

A comparison of data in tables VI and VII shows that for
a given asphaltene content the viscosity of the product oil
depends on the molecular weight of the asphaltenes: FB-40, batch
7, containing 22.3 percent of asphaltenes with a molecular weight
of 459 has a viscosity of 14.3; while FB-38, batch 7, containing
approximately the same amount of asphaltenes but with a molecular
weight of 671 has a viscosity of 85.6.

The increase in viscosity of the product oil during a run
is due primarily to an increase in asphaltenes and possibly
pyridine sols content and not to a change in the molecular weight
of the asphaltenes or to an increase in the viscosity of the
heavy oil. This may be seen by comparing batches 7 and 88 of
run FB-40 in tables VII and VIII. The viscosity of the centrifuged liquid product increases from 59 to 1,433 with an increase
in asphaltene content from 22 to 40 and an increase in pyridine
sols content from 2 to 5. In contrast, the viscosity of the
heavy oil increases only from 12 to 17. There is no increase
but rather a slight decrease in the molecular weight of asphaltenes from 459 to 416.

The increase in viscosity of centrifuged liquid product
with asphaltene content is shown in figure 4, where viscosities
for runs FB-39 and FB-40 are plotted against asphaltene content.

Abstract

In the Energy Research and Development Administration's
SYNTHOIL process, slurries of coal in recycle oil are hydrotreated
on Co-Mo/SiO$_2$-Al$_2$O$_3$ catalyst in turbulent flow, packed-bed reactors. The reaction is conducted at 2,000 to 4,000 psi and
about 450° C under which conditions coal is converted to low-
sulfur liquid hydrocarbons and sulfur is eliminated as H$_2$S.

Product oils from SYNTHOIL runs carried out at 415° and
450° C and 2,000 and 4,000 psi H$_2$ pressures were analyzed with
respect to asphaltene and oil content, elementary compositions
(C, H, S, N), ash and physical properties (specific gravity and
viscosity). Asphaltenes exert a large effect on the viscosity
of the product oil, the viscosity increasing exponentially with
asphaltene content. Viscosity of product oil is not only dependent on the amount but also on the molecular weight of asphaltenes
present. At 415° C, asphaltenes with a molecular weight of 670
are formed and at 450° C asphaltenes with a molecular weight
of 460.

Product oil containing 24 percent asphaltenes of 670 molecular weight has a viscosity of 86 (SSF at 180° F, ASTM D88),
while product oil containing almost the same amount of asphaltenes
(22 percent) but with a molecular weight of 460 has a viscosity
of only 14. Benzene insolubles, heretofore regarded as unreacted
coal, were found to be soluble in pyridine and to exert a large
effect on viscosity.

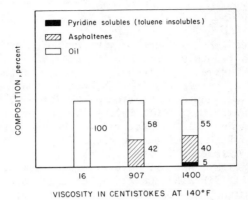

Figure 3. Effect of pyridine solubles (toluene insolubles) and of asphaltenes on viscosity

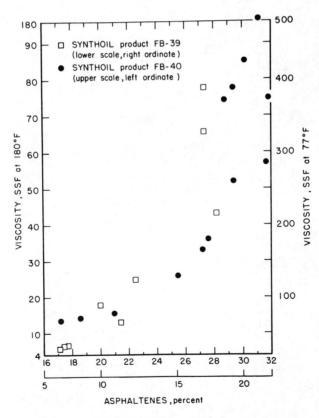

Figure 4. Viscosity of CLP vs. asphaltene content

Literature Cited

1. Akhtar, Sayeed, Lacey, James J., Weintraub, Murray, Reznik, Alan A., and Yavorsky, Paul M. The SYNTHOIL Process--Material Balance and Thermal Efficiency. Presented at the 67th Annual AIChE Meeting, December 1-5, 1974, Washington, D.C.
2. Akhtar, Sayeed, Mazzocco, Nestor J., Weintraub, Murray, and Yavorsky, Paul M. SYNTHOIL Process for Converting Coal to Nonpolluting Fuel Oil. Presented at the 4th Synthetic Fuels from Coal Conference of the Oklahoma State University, Stillwater, Oklahoma, May 6-7, 1974.
3. Yavorsky, Paul M., Akhtar, Sayeed, and Friedman, Sam. Process Developments: Fixed-Bed Catalysis of Coal to Fuel Oil. AIChE Symposium Series, v. 70, No. 137, pp. 101-105, 1974.

8

Coal Liquifaction by Rapid Gas Phase Hydrogenation

MEYER STEINBERG and PETER FALLON

Department of Applied Science, Brookhaven National Laboratory, Upton, N. Y. 11973

The increased awareness in this nation's dependence on foreign sources of crude oil has generated much interest in developing and utilizing U.S. domestic energy resources. Since coal is in such great abundance, it is of particular interest to find an economical method of converting it to a low sulfur alternate fuel capable of being transported through existing pipe lines and used in existing equipment. This would have the three-fold effect of 1) revitalizing the coal industry, 2) providing a greater domestic supply of fuel to the power industry, and 3) increasing the availability of feedstock to the petrochemical industry. A number of coal gasification processes have been developed to the point where large scale application has become feasible. The development of the liquifaction of coal has lagged and when taken together with the advantages of handling, transporting, and higher conversion efficiency, the incentive to perform additional work in converting coal to liquid products becomes apparent.

The technology for liquid phase coal hydrogenation has been in existence for a number of years. (1) For example, during World War II Germany produced much of its liquid fuel via the Bergius Process. This involves forming a slurry of coal, oil and catalyst and heating it under high pressure (10,000 psi) for up to an hour while purging the mixture with hydrogen. Being essentially a batch process involving the solid, liquid, and gas phase, a more direct less stringent method of hydrogenation was desired. In 1962, Schroeder (2) described a process whereby dry particles of coal entrained in a stream of hydrogen at total pressures in the range of from 500 to 6000 psi are rapidly heated to temperatures in the range of 600 to 1000°C; the resulting stream containing the organic products is then quickly cooled to below the reaction temperature and the products separated. This system is claimed to utilize less hydrogen than the liquid phase method because the liquids produced are primarily unsaturated aromatics rather than saturated paraffins and cycloparaffins. Conversion of almost 50% of the coal (on a moisture and ash free basis) to liquid products, was obtained at pressures above 2000 psi, and temperatures in the order of 500°C, and when a molybdenum catalyst was used. Outside this patent literature there appears to be no further information on this system. Moseley and Paterson (3) performed experiments similar to Schroeder but were primarily interested in the production of methane. The fact that they did not report the production of any liquid product is probably due primarily to the higher temperatures utilized, 900-925°C.

Qader, et al., (4) reported work on hydrogenation in a dilute phase free fall reactor at temperatures in the order of 515°C, pressures of 2000 psi and with a heavy dose of catalyst, 15% stannous chloride by weight of coal. Up to 75% conversion was reported with a product distribution of 43% oil, 32% gas and 25% char. The residence time of the coal feed particles was estimated to be in the order of seconds, however, no measurement was made and aromatics were reported after further hydrorefining in a second stage hydrogenation.

More recently, Yavorsky, et al., (5) have reported on the development of a liquid phase (solvent) hydrogenation of coal in a highly turbulent tubular reactor in the presence of a packed solid catalyst. Residence times in the order of several minutes were reported and an oil product is formed. This system is a considerable improvement over the Bergius Process since reduced pressure, in the order of 4000 psi and lower temperature of 450°C are employed. From this literature it thus appears that coal can be converted to liquid and gaseous hydrocarbon products in substantial yields by means of an elevated temperature (450-950°C) high pressure (2000-6000 psi) contact of coal with hydrogen gas. Liquid hydrocarbons are favored in the lower temperature range and shorter contact time (in the order of seconds), while gaseous hydrocarbons increase in yield at the higher temperatures and longer residence times. The basic mechanism appears to involve a thermally induced opening of the polymeric aromatic hydrocarbon structure in coal, allowing hydrogenation to occur in a hydrogen atmosphere, followed by a rapid removal of the liquid and gaseous hydrocarbon products formed so that either further dehydrogenation, repolymerization and carbonization is prevented or further excessive hydrogenation is minimized.

The purpose of the present experiments is to substantiate previous work and to gather additional process and kinetic information especially in a non-liquid non-catalyzed gas phase hydrogenation system. Special effort was made to obtain overall material balances in determining yields.

After a few preliminary runs using a caking eastern bituminous coal and several reactor configurations, it was found that better results could be obtained with a non-caking coal in an entrained down-flow tubular reactor. The coal was dropped down into a tubular reactor through which hydrogen was passed down-flow and entrained and carried the coal down through the heated tube.

Description of Equipment, and Experimental and Analytic Procedures

The experimental equipment (Figure 1) was designed to rapidly heat coal in the presence of hydrogen at elevated temperature and pressure, to maintain these conditions long enough for hydrogenation to take place, to cool the products preventing further reaction and finally to separate and collect the products. Hydrogen is supplied from standard 200 cubic foot cylinders and the operating pressure is controlled by a 4000 psi pressure regulator. Argon is supplied at low pressure to flush the system before and after each run and to activate the normally closed pneumatic valve which shuts the hydrogen supply off in case of sudden loss of pressure. The hydrogen can then pass through either or both of two 5000 psi service rotameters. The gas from one meter goes to the top of the reactor through an electrically heated preheater constructed from a ten foot length of 1/4 inch tubing formed into a 6 inch diameter coil. Since heating is by direct connection to a variable voltage transformer, one side of the system is electrically isolated from the other. Gas from the other meter is used to cool the line between the reactor and the coal feeder and enters just below the feeder. Experience has shown that even with a non-caking coal, this section will become plugged if the temperature is not controlled. This is probably caused by a combination of reduced cross section, small particle size and devolatilization of the coal.

The reactor column is fabricated from an eight foot length of 1 in. O.D. x 0.120 in. wall thickness type 304 stainless steel tubing. The coal feeder is a pressurized 40 gram capacity hopper having an orifice in the base and a tapered plug operated by an electromagnet acting through a non-magnetic stainless steel wall of the hopper. The feed rate is controlled by both a pulse generator operating an electronic relay which actuates the solenoid and by controlling the distance the tapered plug is raised off the seat during its vertical cycling.

A North Dakota lignite and a New Mexico sub-bituminous coal were mainly used in these experiments. Coal preparation consisted of size reduction and drying. Ideally, the particle size should fall somewhere between the largest size with which the maximum rate of hydrogenation takes place and a size large enough so that inter-particle attraction and agglomeration does not occur. With the ball mill used in these experiments for size reduction, a large quantity of fines were produced when grinding to a maximum particle size of ≤ 50 or $\leq 150 \, \mu$. The inside of the feeder had to be coated with a fluorocarbon release agent which was baked in place. The coal was magnetically agitated by pieces of a finned rod attached to the tapered plug to prevent aggregation and clogging of the feeder.

To prevent oxidation all grinding was conducted in an inert atmosphere. Drying in air at temperatures slightly above $100^{\circ}C$ was also found to cause oxidation and for this reason drying and storage was performed in a vacuum oven.

Analysis of the coal used in these experiments and of a typical char from the lignite coal produced during a series of runs through the hydrogenator are shown in Table 1. The analyses are on a moisture-free basis and were obtained by the U.S. Bureau of Mines using standard ASTM analytical procedures.

After the coal passed through the reactor, the char and any unreacted coal were collected in an air cooled tubular trap containing a stainless steel wool filter attached to the base of the column. The quantity collected is determined by weighing the trap before and after each run.

The products pass through the trap along with the excess hydrogen which is still at elevated temperature and reactor pressure. These gases then pass through a double U-tube water cooled trap (25-30°C), which removes any oils produced and most of the moisture. The oil deposits on the walls of the trap and due to the relatively small quantity produced and the large trap surface area, collection becomes difficult. By cleaning with a cotton swab, the oil collected appears to be light in color and relatively low in viscosity, similar to a motor oil. The quantity produced is determined by weighing the trap before and after the run. The gases are then passed through a metering valve reducing the pressure to atmospheric. The remaining liquids contained in the gases are collected in alcohol-dry ice cooled glass trap (-72°C). Samples of the hydrocarbons collected in this trap have consistently been analyzed to contain over 98% benzene, the remainder being toluene with a trace of xylene. The quantity produced is determined by weighing the trap before and after the run and by volume in a small graduate attached to the glass trap. Water from the reactor usually collect in this trap and appears as a separate layer in the graduate. This water and that collected in the U-tube trap have been analyzed on a Beckman Carbon Analyzer and shows some insignificant amount of dissolved hydrocarbons. The liquid hydrocarbons collected in this trap are analyzed by gas chromatography using a column of 50-80 mesh Porapak Q and a flame ironization detector. The chromatograph is calibrated to detect benzene, toluene, xylene ($\geq C_6$ liquids) and all the light hydrocarbon gases, methane, ethane, propane, hexane and pentane ($\leq C_5$ gases). The remaining gas is then vented to the atmosphere. At about two-thirds of the time period through the run, a 300 cc gas sample is taken in a glass sampling vessel inserted in the vent line. Due to the limited capacity of the coal feed hopper, the duration of most runs was in the order of 10 to 15 minutes. Steady state operation is usually obtained quickly since in about one or two minutes at the flow rates used several reactor volume changes have occurred. The gas sample is analyzed in the same manner as the liquid sample mentioned above. Aside from excess hydrogen, the gas sample usually contains mostly methane and ethane, the methane being present in approximately twice the abundance as the ethane. The next greatest constituent is usually benzene in the vapor phase. There appears to be very little C_3, C_4, and C_5 hydrocarbons present. Periodically these same samples

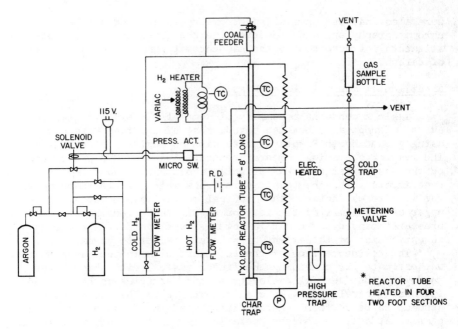

Figure 1.   Coal hydrogenation tubular reaction experiment

TABLE 1

COAL AND CHAR FEED ANALYSIS

% by wt - Moisture Free Basis

	North Dakota Lignite	New Mexico Sub Bituminous	Char from Lignite
Hydrogen	4.29%	4.86%	2.74%
Carbon	62.39	64.21	73.63
Nitrogen	0.97	1.41	0.85
Oxygen	21.75	12.26	0.37
Sulphur	1.17	0.61	1.42
Ash	9.43	16.65	20.99
	100.00	100.00	100.00

were also analyzed by mass spectrometry as a check on the gas
chromatography and also to determine the concentration of con-
stituents not detected by the gas chromatograph, such as oxides
of carbon.

Experimental and Calculated Results

Table 2 contains a summary of some of the experimental results. Except as otherwise noted, most of the runs were made
using ground North Dakota lignite with no addition of catalyst.
The reactor operating temperature condition of 700°C was chosen
on the basis that this was slightly above 650°C found in previous
experiments to be the temperature below which little reaction with
the lignite was noted. The 1500 psi operating pressure is the
upper safety limit of the reactor at the 700°C temperature. Lower
pressures tended to result in lower liquid yields. A thorough
investigation of the effect of pressure is yet to be made.

The residence time of the coal particles in the reaction zone
was estimated by combining the linear gas velocity through the
tube with the free fall velocity of the particles as calculated
from data correlated by Zenz and Othmer (6). The particles were
assumed to be uniform spheres distributed in a highly dilute gas
phase. At a fixed reactor operating temperature and pressure, the
residence time becomes dependent upon the particle size and the
gas flow rate. Most of the original experiments were conducted
using ≤ 50 micron coal particles. In an attempt to improve the
uniformity of the coal feed flow and decrease the residence time,
this was increased to ≤ 150 microns. As shown in Table 3, it was
not possible with the frequent sieving and grinding method used
to reduce the presence of significant amounts of ≤ 50 micron particles.

Using Figure 2, which is a plot of residence time as a function of coal particle size based on the Zenz and Othmer correlation, it can be shown that due to the relatively large quantity
of fines present, increasing the maximum size by a factor of three
only reduces the average residence time by about 30%. Figure 2
also shows the effect of reducing residence time by increasing the
hydrogen flow rate from the minimum used in these experiments
(0.7 gm/min) to the maximum (4.5 gm/min).

Brown and Essenhigh (7) have suggested that the free fall
velocity may be much greater than that calculated here. In their
experiments using cork particles falling through an air filled
tube, the particle cloud acted as a porous plug, falling at a
rate of approximately 57 ft/min. Using the Zenz and Othmer correlation, this velocity is calculated to be only 17 ft/min. Thus,
the coal residence time may be much shorter than determined from
Figure 2. Additional experimental effort is required to measure
the actual residence in the system.

The hydrogen fed to the system is much in excess of that
utilized in the hydrogenation reaction. The net use of hydrogen

TABLE 2

HYDROGENATION OF COAL IN AN ENTRAINED TUBULAR REACTOR

Coal - North Dakota Lignite

Reactor - 0.75" I.D. x 8 ft Long Stainless Steel

Run No.	5/3	5/21	6/3	6/7	6/14	7/30	9/9	9/13	9/16	11/21	9/11
H_2 Pressure (psi)	1000	1400	1500	1500	1500	1500	1500	1500	1500	1500[3]	1500
Reactor Temp (°C)	700	700	700	700	700	700	700	700	700	700	700
H_2 Preheat Temp (°C)	350	350	350	350	R.T.	300	600	600	600	R.T.	600[4]
Coal Rate (gm/min)	1.5	0.78[1]	1.47	1.66	2.93	1.63	2.53	1.4	3.7[2]	1.08	0.94[4]
H_2 Rate (gm/min)	0.7	0.74	4.14	4.14	2.5	1.6	1.75	1.75	2.1	21.3[3]	1.75
H_2 Linear Velocity Ft/Sec	0.056	0.059	0.33	0.33	0.20	0.128	0.14	0.14	0.168	0.086	0.14
Max.Coal Particle Size (Microns)	≤50	≤50	≤50	≤50	≤50	≤150	≤150	≤150	≤150	≤150	≤150
Est. Max. Coal Residence Time (Sec)	38	36	16	16	22	20	18	18	16	30	18
Duration of Run (Min)	10	15	20	17.5	10	11.5	10	10	10	10	10
Raw Carbon Balance - % C Converted											
% C as Liquid HC ($\geq C_6$)	8.4	7.3	35.8	21.2	17.5	25.0	19.1	23.8	12.9	4.0	0
Gaseous HC (C_1-C_5)	19.6	34.3	45.7	10.6	19.5	5.3	1.7	26.8	30.6	7.7	0
C Oxides[5]	6.0	6.0	6.0	6.0	6.0	6.0	6.0	6.0	6.0	16.2	-
Unreacted	49.5	48.3	34.1	32.9	44.0	41.9	27.9	42.9	44.1	68.8	95.3
% Total Conversion of C to Gas & Liquid	28.0	41.6	81.5	31.8	37.0	30.3	20.8	50.6	43.5	11.7	0
% C Accounted for	83.5	95.9	121.6	70.7	87.0	78.2	54.7	99.5	93.6	96.7	95.3
Modified C Balance - % C Converted											
% C as Liquid HC ($\geq C_6$)	8.4	7.3	35.8	21.2	17.5	25.0	19.1	23.8	12.9	4.0	0
Gaseous HC (C_1-C_5)	36.1	38.4	24.1	39.9	32.5	27.1	47.0	27.3	37.0	11.0	4.7
C Oxides[5]	6.0	6.0	6.0	6.0	6.0	6.0	6.0	6.0	6.0	16.2	-
Unreacted	49.5	48.3	34.1	32.9	44.0	41.9	27.9	42.9	44.1	68.8	95.3
% Total Conversion of C to Gas & Liquid	44.5	45.7	59.9	61.1	50.0	52.1	66.1	51.1	49.9	15.0	4.7

(1) 1% Ammonium Molybdate Added
(2) New Mexico Sub-Bituminous Coal
(3) Argon Used in Place of Hydrogen
(4) Char Collected from Previous Lignite Coal Runs Used in Place of Coal as Feed Material to the Tubular Reactor
(5) For Hydrogen Runs, Analysis Indicates 1% CO_2 and 5% CO
     For Argon Run, Analysis Indicates 11.5% CO_2 and 4.7 % CO

TABLE 3

COAL PARTICLE SIZE DISTRIBUTION

Particle Size (Microns)	Weight Percent
150 - 125	4.8
125 - 90	21.4
90 - 50	21.0
≤ 50	52.8

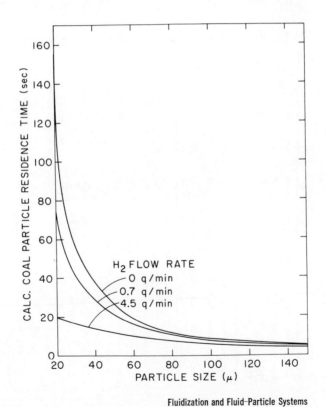

Fluidization and Fluid-Particle Systems

Figure 2. Coal particle residence time vs. particle size.
H_2 at 1500 psi + 700°C.

for a high yield run where, for example, 60% of the available
carbon from the lignite is hydrogenated into equal parts of liquid
and gas, is only about 5 grams hydrogen per 100 grams of coal.
An additional 50% of this amount of hydrogen, originates from the
coal itself to make up the hydrogen balance in the hydrocarbon
products. Hydrogen for reaction with the sulfur, nitrogen and
oxygen contained in the coal is taken into account for this balance. It should be noted that the lignite has an especially high
oxygen content. The large excess hydrogen feed serves the purpose
of transferring heat to the coal, transporting the products out of
the reactor and producing variations in the coal residence time by
changes in hydrogen velocity.

The carbon balance given in Table 2 was made by weighing the
liquids and char collected in each trap, determining analytically
or, as in the case of the light oil, estimating the carbon content
and analysis of a sample of the effluent gas. The oil was assumed
to contain some amount of aromatic with approximately 90% carbon
content. If this oil was primarily aliphatic, the carbon content
would be closer to 85%. This would only affect the overall balance by a small amount since the oils make up approximately 40%
of the liquids collected, the rest being chromatographically
analyzed as benzene, toluene and xylene (BTX) with the major fraction ($\geq$ 98%) being benzene.

As mentioned previously, a sample of the effluent gas is
analyzed chromatographically for light hydrocarbons (C_1-C_5). Only
a small fraction of higher hydrocarbons ($\geq C_6$) is found in the gas
sample and this is added to the total liquid formed. The total
carbon content of the gas is obtained from the results of this
analysis, the gas flow rate and the duration of the run. Typical
results of mass spectrometric analyses also indicate that about
5% of the available carbon appears as carbon monoxide and 1% as
carbon dioxide in the effluent gas. As shown in Table 2, most of
the experiments indicate over 70% of the carbon can be accounted
for by direct analysis. Since the amounts of coal used and con-
densed liquid and char produced were measured gravimetrically and
there was no holdup in the system, any inability to account for
all the carbon fed to the reactor must be attributed to loss in
the effluent gas. The carbon balance was therefore modified by
assuming the missing carbon to be in the form of a light hydrocarbon in the effluent gas. This modified carbon balance is given
in the lower half of Table 2.

Figure 3 shows the effect of coal particle residence time on
the fraction of carbon converted to liquid and gaseous products.
Based on this correlation, the production of gaseous products
appears to be less sensitive to coal residence time than the production of liquids. It also appears that the production of liquids is favored by shorter residence time.

Figure 4 gives a plot of the conversion to liquids as a
function of coal feed rate for the shorter coal residence times
(16-22 sec). This figure indicates a trend towards lower liquid

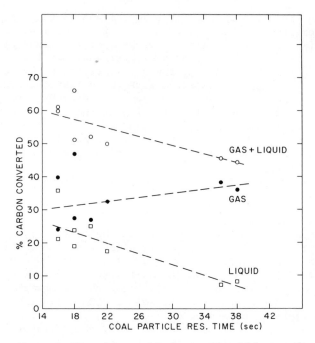

Figure 3. Gas phase coal hydrogenation yield vs. coal residence time. North Dakota lignite—reactor, 0.75" i.d. × 8 ft; pressure, 1000–1500 psi; temperature, 700°C; particle size, ≤ 150 μ.

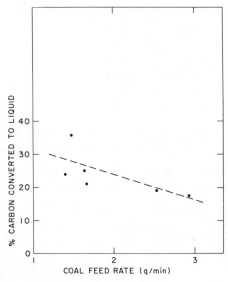

Figure 4. Gas phase coal hydrogenation liquid yield vs. coal feed rate. North Dakota lignite—reactor, 0.75" i.d. × 8 ft; pressure, 1500 psi; temperature, 700°C; particle size, ≤ 150 μ; residence time, 16–22 sec.

conversion at higher coal feed rates.. Even at the higher feed
rates, the coal is in a very dilute phase in the gas, so that
there should be no tendency towards a coal saturation effect in
the system. Since much of the heat needed to bring the coal to
reaction temperature can be obtained as radiant heat from the reactor wall, this effect of coal feed rate could be attributed to
a shadowing effect, reducing the heating rate at higher coal feed
rates. This assumes that rapid heating of the coal particles is
needed for optimum liquid yield. From experience in running the
equipment, it was also found that while preheating the hydrogen
increased the overall yield, the difference in preheating between
350°C and 600°C did not appear appreciable.

Several runs were made with the addition of 1% ammonium
molybdate to the lignite for purposes of testing catalytic effects. Under similar flow conditions there appears to be little
effect on the yields.

In the run made with a New Mexico sub-bituminous coal, an
appreciable quantity of gaseous hydrocarbons was produced. The
main difference between the New Mexico coal and the North Dakota
lignite is that the sub-bituminous coal has a lower oxygen content and a higher ash content.

Runs were also made using an inert gas, argon, in place of
hydrogen at the similar pressure and linear velocity through the
reactor. The results of a typical one of these runs is given in
Table 2. Although almost 11 grams of coal was fed, no measurable
quantity of product was found in the high pressure water cooled
trap, and only a few drops of what was predominantly benzene was
collected in the alcohol-dry ice cooled trap. Analysis of the
gas sample showed methane and about one-fifth as much ethane. The
char contained a higher carbon content (77%) than when hydrogen
was used (72%). Almost 70% of the available carbon remained unreacted. The production of oxides of carbon was much higher with
argon (16%) than for a typical hydrogen run (6%). Also, the CO
to CO_2 ratio was much higher (5 times) with hydrogen than with
argon (2 times). These results indicate that volatilization of
lignite in an inert gas stream produces much less gaseous and
liquid hydrocarbon products (15% of C) under similar conditions
of pressure, temperature and residence time than when hydrogen is
present (45 to 66% of C) and thus appreciable hydrogenation is
occurring.

A test was made to determine the reactivity of the char remaining after a hydrogenation run. For this purpose, char collected from a series of previous runs was recycled through the
system. Analysis of this char appears in Table 1. Over 95% of
the material fed was recovered as a solid in the first trap. No
liquid was produced and only a trace of light hydrocarbons were
found in the gas. Similar results were obtained when feeding a
coconut charcoal to the system. There thus appears to be little
reactivity towards hydrogen in the char produced from an original
lignite hydrogenation run.

Conclusions

The objectives of these hydrogenation experiments were mainly directed toward producing the largest quantity of liquid product and reducing the unreacted carbon remaining. Although frag-
mented, this experimental study indicates that significant yields
of liquid and gaseous hydrocarbon products can be obtained by an
elevated temperature and pressure gas phase hydrogenation of
North Dakota lignite without the addition of a catalyst. Based
on a modified carbon balance, a range of yields obtained in a
number of experimental runs under the operating conditions indicated can be listed as follows:

Pressure	1500 psi
Temperature	700°C
Coal particle size	≤ 150 μ
Coal residence time	16 to 22 sec.

Yield as % conversion of carbon in lignite

Liquids ($\geq C_6$)	18-25% ($\sim$60% C_6H_6)
Gas ($\leq C_5$)	27-40 (mainly CH_4 & C_2H_6)
Oxides of C	6
Unreacted C	33-44

Several general trends were noted. The yield of liquid products is strongly favored by shorter coal residence times, while
the yield of gaseous hydrocarbons increases only slightly with
increases in coal residence times. At shorter residence times,
increasing the coal feed rate decreases the yield of liquid products.

Volatilization of lignite in an inert gas stream under similar flow conditions produced much smaller yields of liquid and
gaseous hydrocarbon products. Addition of a catalyst caused no
appreciable effects on the yield.

The results of this study should be considered as work in
progress. Much additional work needs to be performed to refine
the experimental procedure and to obtain additional information
and a fuller understanding of the results.

Acknowledgement

The authors would like to thank Mr. Gerald Farber
for the construction and operation of the equipment,
the U.S. Bureau of Mines for the coal analysis and
Robert Smol, Joe Forrest, and Robert Doering of Brookhaven National Laboratory for the gaseous and liquid
product analysis. Acknowledgement is also made to
the group at City University of New York, Arthur Squires,
Robert Graaff, and Sam Dobner for their helpful dis-
cussions on rapid reactions of coal with hydrogen.

Abstract

The short contact time, high temperature, non-catalyzed gas phase reaction of hydrogen with coal in a tubular reactor has been investigated. A North Dakota lignite coal was hydrogenated by entraining fine coal particles ($\leq$ 150 μ) in a hydrogen stream through a 0.75" I.D. by 8 ft long tubular reactor at 700°C and 1000 to 1500 psi. The coal flow rate varied between 0.8 and 3.7 gm/min. and the hydrogen flow rate varied between 0.7 and 4.1 gm/min. Residence times were calculated to vary between 16 and 38 seconds. The unreacted carbon was collected in an air cooled trap, the light oils in a water cooled trap on the high pressure side of the system and the lighter liquids consisting predominantly of benzene with smaller amounts of xylene and toluene (BTX) were collected in an alcohol-dry ice cooled trap after reducing the pressure to one atmosphere for venting. The exit gas was analyzed. Based on a modified carbon balance, the yield, or conversion of carbon in the feed coal, to liquid products ($\geq C_6$) usually varied between 18 and 25%, approximately 60% of which is benzene. The gases ($\leq C_5$) varied between 27 and 40% consisting mainly of methane and ethane, and the oxides of carbon were 6% of the available carbon feed in the lignite. A trend indicating higher liquid yields at shorter coal residence times was noted. A decrease in liquid yield at increased coal feed rate when the residence time remained approximately constant was also noted.

Volatilization of lignite in an inert gas stream under similar flow conditions produced much smaller yields of liquid and gaseous hydrocarbon products. Addition of a catalyst caused no appreciable effects on the yield. Much additional work is required to obtain a fuller understanding of the results obtained.

Literature Cited

1. Mills, G. Alex, Ind & Eng. Chem. (1969), 61, 6-17.
2. Schroeder, W. C. (assigned to Fossil Fuels, Inc.), U.S. Pat. 3,030,298 (April 17, 1962) and U.S. Pat. 3,152,063 (October 6, 1964).
3. Moseley, F. and Paterson, D., J. Inst. Fuel (November 1967), 523-530.
4. Qader, S. A., Haddadin, R. A., Anderson, L. L., and Hill, G. R., Hydrocarbon Processing (September 1969) 48, 147-152.
5. Yavorsky, P., Akktar, S., and Friedman, S., Chem. Eng. Progress (1973) 69, 51-2.
6. Zenz, F. A. and Othmer, D. F., Fluidization and Fluid-Particle Systems, p 236, Reinhold Publishing, New York, 1960.
7. Brown, K. C. and Essenhigh, R. H., Dust Explosions in Factories: A New Vertical Tube Test Apparatus, Safety in Mines Research Establishment Research Report, No 165, Ministry of Power, Sheffield, England (April 1959).

9

Consideration of Catalyst Pore Structure and Asphaltenic Sulfur in the Desulfurization of Resids

RYDEN L. RICHARDSON and STARLING K. ALLEY

Union Oil Co. of California, Union Research Center, Brea, Calif. 92621

The residuum fraction of a full range crude is that fraction remaining after all of the distillate is taken overhead. As noted in Figure 1, this residuum fraction or resid may be obtained with either atmospheric or vacuum fractionation, yielding either long or short resid.

Table I shows that, upon distillation of a Kuwait crude into several fractions, sulfur components concentrate mostly in the resid fraction. Upon hydrotreating each fraction at fixed reactor conditions, the resid sulfur is more difficult to remove than the gasoline sulfur or the gas oil sulfur.

CHARACTERIZATION OF RESIDS

The distinguishing features of resid feedstocks are (1) the presence of asphaltenes or pentane-insolubles, (2) high carbon residues, (3) the presence of metals, mainly nickel and vanadium, and (4) unknown endpoints insofar as boiling range is concerned.

Table II compares two atmospheric resids, West Coast and Kuwait, in a traditional manner. The obvious differences include sulfur, nitrogen, asphaltenes, total metals and mid-boiling point. Apart from sulfur content, one might surmise a greater catalyst demand by the West Coast feedstock in that the boxed values suggest heavy coke laydown and metals deposition. Neither of the sulfur values is boxed because there is no indication as to (1) what fraction of the sulfur is refractory or "hard" sulfur, nor (2) the degree of desulfurization to be achieved.

Table III extends the comparison of these resids with an emphasis on reactivity, asphaltene characteristics, compound types and the refractory forms of sulfur, such as benzothiophenes and asphaltenic sulfur.

The lower rate constant of the Kuwait feed indicates its greater refractoriness or resistance toward desulfurization. These second order rate constants were measured at 675°F with a presulfided cobalt moly catalyst having a catalyst age of 10 days. Conversion levels did not exceed 75% and hence did not go deeply

TABLE I

FRACTIONATION AND HYDROTREATING

FRACTION	BOILING RANGE, °F	VOL. %	SULFUR WT. %	HYDROTREATED PRODUCT S, %	SULFUR REMOVAL, %
GASOLINE	X – 400	25	0.5	0.02*	96
GAS OIL	400 – 650	25	1.8	0.5 *	72
LONG RESID	650 +	50	3.7	1.6 *	57
SHORT RESID	1050 +	(20)	(5)	– – –	– –

* 780 °F, 1000 psig, 1 LHSV, 4000 SCF/B

TABLE II

COMPARISON OF WEST COAST AND KUWAIT
ATMOSPHERIC RESIDS

FEED DESIGNATION	WEST COAST	KUWAIT
GRAVITY, °API	11.5	17.4
DISTILLATION, °F		
IBP/10	536/725	447/634
50/60	996/1051	937/1020
SULFUR, %	1.73	3.66
NITROGEN, %	0.90	0.203
ASPHALTENES, %	11.2	6.3
CONRADSON CARBON, %	10.7	8.1
POUR POINT, °F	75	30
METALS, ppm		
COPPER	1	1
IRON	53	2
NICKEL	75	11
VANADIUM	63	38

TABLE III

EXTENDED COMPARISON OF WEST COAST AND KUWAIT RESIDS

FEED DESIGNATION	WEST COAST	KUWAIT
RATE CONSTANT	52	38
TOTAL SULFUR, %	1.73	3.66
ASPHALTENIC SULFUR, %	0.22	0.38
GEL. PERM. CHROMO.		
ASPHALTENE MODE DIAM., Å	60	70
RELATIVE AMOUNT OF ASPHALTENES AT D = 150 Å	0.72	0.78
LIQUID CHROMO		
COMPOUND TYPE		
SATURATES, %	20.6	26.7
AROMATICS, %	41.2	50.4
POLAR AROMATICS, %	22.9	10.4
ASPHALTENES, %	10.7	6.3
FEED DESIGNATION	WEST COAST	KUWAIT
SULFUR CONTENT		
SATURATES, ppm	22	34
AROMATICS, %	0.94	2.36
POLAR AROMATICS, %	0.41	0.59
ASPHALTENES, %	0.22	0.38
HIGH MASS ON X-1100 °F O.H.		
THIOPHENES	2.9	2.3
BENZOTHIOPHENES	7.4	15.0
TRIBENZOTHIOPHENES	0.0	0.0
IR ABSORBANCE		
AROMATIC RINGS	0.43	0.33
ALKYL GROUPS	0.38	0.42
RATIO OF NMR PEAK HEIGHTS		
CH_2/CH_3	LOW	HIGH

into asphaltenic sulfur. Figure 2 shows the GPC curves of the
cited resids and their asphaltene fractions. Note the slight predominance of very large asphaltenes in the Kuwait resid. Also
note that very large molecules exist in the non-asphaltenic fraction of both resids. For the conversion levels under consideration, we regard the large proportion of sterically hindered aromatic sulfur in the Kuwait resid largely responsible for its greater
refractoriness. Returning to Table III, one notes the high values
for aromatic sulfur (2.36 vs. 0.94%) and benzothiophenes (15.0 vs.
7.4%) in the Kuwait feed. Also, the IR and NMR spectra indicated
that the Kuwait aromatics are more highly substituted, the R
groups being naphthenes and/or long paraffins.

A pertinent example of steric hindrance on the desulfurization of thiophenes has been reported by a Gulf investigator (1).
Alkyl or ring addition at the 2 position can reduce the reaction
rate by a factor of 100.

REFRACTORINESS OF ASPHALTENES

When sulfur conversion levels are pushed above 90%, the
unique refractoriness of asphaltenes becomes dominant. The tendency of a fine pore catalyst to partially exclude asphaltenes and
the complete steric hindrance or "burying" of sulfur in asphaltenes contribute to this refractoriness.

However, before considering the fate of asphaltenic sulfur at
high reactor severities, some patent aspects of catalyst pore
structure would be of interest. Table IV shows a divergence of
opinion as to the desirability of asphaltene exclusion.

It is noted that patent (2) favors the exclusion of asphaltenes by maximizing surface area contained by pore diameters of
30-70 Å. Patent (3) discloses an upper limit on the amount of
macropore volume represented by pore diameters greater than 100 Å.
Patents (4) and (5) suggest that catalysts containing mostly micro-
pores will be poisoned soon; asphaltenes penetrating the larger
pores subsequently will block entrance to the smaller pores.
Patent (6) claims the desirability of intermediate pores (100-1000
Å) plus channels (>1000 Å) "to take up preferentially adsorbed
large molecules without causing blockage, so that the smaller size
pores can desulfurize 'smaller' molecules." Patent (7) also pre-
fers the open structure for collection of coke and metals. It
specifies 0.3 cc/g of pore volume in diameters larger than 150 Å
and "many pores from 1,000-50,000 Å."

With this brief consideration of variations in catalyst pore
structure, let us examine the pore structure of two catalysts used
in this refractoriness study. One observes in Table V only slight
differences between the two cobalt moly catalysts, T and R. They
are typified by high surface area, small micropore mode diameters
and low macropore volumes.

Upon processing Kuwait resid over these catalysts, a similar
trend in product distribution is shown in Table VI. Saturates in-

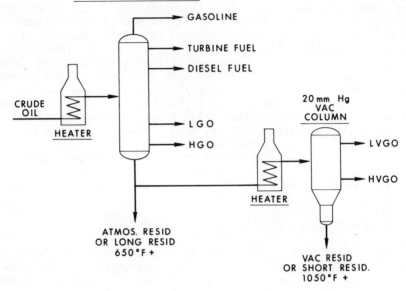

Figure 1.   Distillation of crude oil

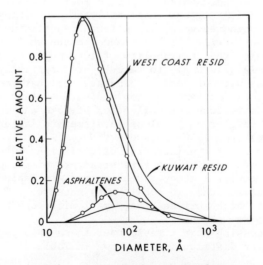

Figure 2.   GPC molecular size distribution

TABLE IV

SOME PATENTS ON PORE STRUCTURE
OF RESID HYDROTREATING CATALYSTS

PATENT CODE NUMBER	CATALYST COMPOSITION	SPECIAL FEATURES	
		PORE VOLUME	PORE DIAMETER
(2)	CoMo, NiMo	— — —	SURFACE AREA (SA) OF 30–70 Å PORES ≥ 100 m^2/g.
(3)	CoMo, Ni Mo, WMo. ALUMINA WITH 1-6 % SILICA.	<0.25 cc/g IN PORES > 100 Å	TOTAL SA > 150 m^2/g. SA OF 30–70 Å PORES ≥ 100 m^2/g.
(4)			APPROX. 85 % OF PORE VOLUME IN 50-200 Å RANGE.

PATENT CODE NUMBER	CATALYST COMPOSITION	SPECIAL FEATURES	
		PORE VOLUME	PORE DIAMETER
(5)	NiCoMo. NO SILICA WITH ALUMINA.	0.46 cc/g	REGULAR DISTRIBUTION 0-240 Å. AVERAGE = 140-180 Å.
(6)	CoMo (+ Ni)	0.45–0.50 cc/g	60–70 Å PLUS 100-1000 Å PLUS CHANNELS > 1000 Å. SA = 260-355 m^2/g.
(7)	CoMo, NiW. NO SILICA WITH ALUMINA	0.3 cc/g IN PORES >150 A	MANY PORES 1,000 – 50,000 Å

TABLE V

PORE STRUCTURE OF COBALT MOLY CATALYSTS
USED IN REFRACTORINESS STUDY

CATALYST DESIGNATION	T	R
SURFACE AREA (SORPT.), m^2/g	284	307
MODE DIAMETER, $\mathring{A}$	67	65
TOTAL PORE VOL., ml/g	0.44	0.52
MACRO (D $\gtrless$ 100 $\mathring{A}$) PORE VOL., ml/g	0.012	0.035

TABLE VI

CHANGE IN PRODUCT COMPOUND TYPE AND
ASPHALTENIC SULFUR CONTENT WITH
CONVERSION LEVEL

REACTOR CONDITIONS	FEED	CATALYST T			CATALYST R		
TEMPERATURE, °F		655	655	709	665	670	710
PRESSURE, psig		800	1292	1292	800	1292	1292
LHSV, VOL./(VOL.) (HR)		0.5	0.3	0.3	0.5	0.3	0.3
CONVERSION LEVEL,		74	83	92	74	85	94
COMPOUND TYPE, %							
SATURATES	30.6	36.8	42.8	43.1	37.5	39.7	45.5
AROMATICS	45.8	47.6	46.8	43.5	47.6	46.4	44.3
POLAR AROMATICS	17.3	10.7	6.2	10.3	10.8	11.3	8.2
ASPHALTENES	6.3	4.9	4.2	3.1	4.1	2.6	2.0
TOTAL SULFUR, %	3.66	0.95	0.63	0.29	0.96	0.55	0.23
ABSOLUTE ASPHALTENIC SULFUR, %	0.45	0.32	0.26	0.18	0.24	0.15	0.09
RELATIVE ASPHALTENIC SULFUR, %	12	34	41	62	25	27	39

crease, aromatics remain fairly constant and both polar aromatics and asphaltenes decrease. Adsorption chromatography is the basis for type separation: pentane elutes the saturates, ether elutes the aromatics, and benzene-methanol elutes the polar aromatics. Quite clearly, the relative amount of asphaltenic sulfur increases as the product total sulfur decreases. This is depicted in Figure 3.

Whereas the relative amount of aromatics remained fairly constant as sulfur conversion level was increased to 92-94%, the relative amount of sulfur in the aromatic fraction decreased markedly. This also is depicted in Figure 3. Polar aromatics are intermediate to the aromatics and asphaltenes in regard to this behavior.

These different distributions of sulfur with conversion level resemble those reported by Drushel for Safaniya resid (8).

The removal of metals with the asphaltenic sulfur is observed in Figure 4. This response is consistent with an asphaltene model in which vanadium and nickel are buried as porphyrins or sandwich compounds (9). The slightly higher removal of vanadium reflects a general tendency for vanadium to deposit on the catalyst more readily than nickel.

EFFECT OF COKE DEPOSITION ON PORE SIZE DISTRIBUTION

A first step toward catalyst design is to relate pore structure roughly to asphaltene dimensions. Another step is to consider pore structure of the used catalyst and the probable size of asphaltenes at reactor temperature.

Figure 5 is a histogram showing the distribution of pore volume vs. pore diameter for alumina carrier, fresh cobalt molybdenum catalyst and used cobalt molybdenum catalyst. There was a slight change in mode diameter when the carrier was loaded with about 20% active metal oxides. The pore volume was reduced from 0.60 to 0.53 ml/g. However, accumulation of about 17% coke during the processing of West Coast resid greatly shifted the mode downward and reduced the total pore volume from 0.53 to 0.30 ml/g. (All of these pore volumes have been normalized to 1.0 gram of alumina).

Additional comparisons of fresh vs. used catalyst are shown in Table VII. The coke laydown occurred mainly in the micropore region, causing a substantial loss of surface area. Coke laydown also was observed in the macropore region of the bimodal catalyst shown at the bottom of the table.

ASPHALTENE EXCLUSION AS A FUNCTION OF TEMPERATURE

The exclusion of asphaltenes first was approached by the simple method of immersing a given amount of catalyst in a volume of resid equal to 3 times the pore volume of the catalyst. If asphaltene exclusion were to occur to a significant degree, then there

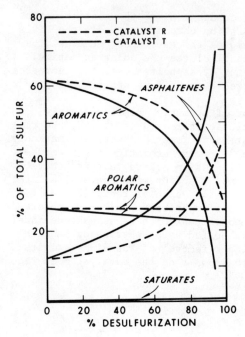

Figure 3.   Distribution of sulfur with conversion level

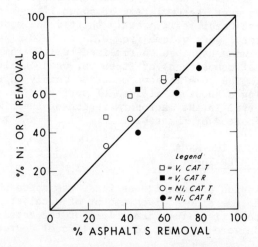

Figure 4.   Removal of metals vs. asphaltenic sulfur

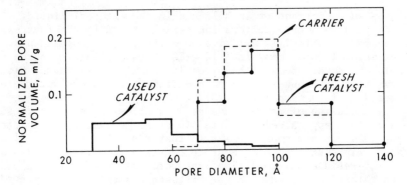

Figure 5. Histograms of pore volumes vs. pore diameter for carrier, fresh catalyst, and used catalyst

TABLE VII

CHANGE IN PORE STRUCTURE
DUE TO COKE DEPOSITION

SAMPLE DESCRIPTION AND DESIGNATION	PORE VOLUME, ml/g			MODE DIAMETER Å	SURFACE AREA m²/g
	MACRO	MICRO	TOTAL		
S − 0293 − A	0.03	0.36	0.39	90	207
USED S − 0293 − A	0.04	0.13	0.17	55	164
V − 8477 − B	0.02	0.35	0.37	85	136
USED V − 8477 − B	0.02	0.15	0.17	60	102
X − 0234 − A	0.02	0.41	0.43	90	167
USED X − 0234 − A	0.02	0.21	0.23	60	140
H − 6013	0.70	0.45	1.15	95, 350	350
USED H − 6013	0.39	0.30	0.69	80, 400	174

would be an enrichment of asphaltenes in the external liquid.

Using a 24-hour equilibration period of 212°F, the four samples shown in Table VIII all give an enrichment of asphaltenes. The exterior liquids contained 12.0-13.8% asphaltenes compared to 6.3% in the Kuwait feed. These results qualitatively agree with observations of Drushel who used a different technique with Safaniya resid at room temperature (8).

The exclusion of asphaltenes as a function of temperature subsequently was approached with the aid of gel permeation chromotography (GPC) in a manner similar to that described by Drushel. Catalyst was equilibrated 4 hour at constant temperatures with a volume of resid equal to 3 times the pore volume of the catalyst. During this time, the system was nitrogen blanketed and agitated every 1/2 hour. The external liquid subsequently was drained through a screen and submitted for GPC analyses and metals content. The internal liquid was extracted from the catalyst pores first by benzene and then 50/50 methanol-benzene. After evaporation of these solvents, the internal liquid was submitted for identical analyses.

The GPC curves in Figure 6 show asphaltene or "large molecule" exclusion at the nominal temperatures of 200, 400 and 600°F. The effect decreases with increasing temperature: planimeter areas fall in the ratio of 1.00, 0.65, and 0.40.

Metals distribution, which in some degree should relate to asphaltene distribution, are shown for the actual equilibration temperatures in Table IX. Nickel showed the expected enrichment at all temperatures. Vanadium responded similarly except at 403 and 603°F. Progressive sulfiding (0.6, 0.7, 1.4% S) and coke laydown (0.8, 1.5, 7.8% C) were observed for the used catalysts and hence represent an experiment complication.

CONCLUSIONS

The points to be emphasized include the following:
1. The refractoriness of resid feeds has been considered from the standpoint of process severity and division between "hard" and "easy" sulfur.
2. For lower conversions levels, where asphaltenic sulfur removal is not deep, refractoriness appears to be largely influenced by a large proportion of sterically hindered aromatic sulfur compounds. Substituted thiophenes, benzothiophines and dibenzothiophenes are representative compounds.
3. Asphaltenic sulfur is the most refractory specie in resids and the removal of metals, particularly nickel, correlates well with removal of asphaltenic sulfur.
4. Coke deposition alters catalyst pore size distributions significantly and is an effect to be followed in regard to catalyst aging.
5. The exclusion of asphaltenes in Kuwait resid is observed by an autoclave technique at 212°F with several catalysts having

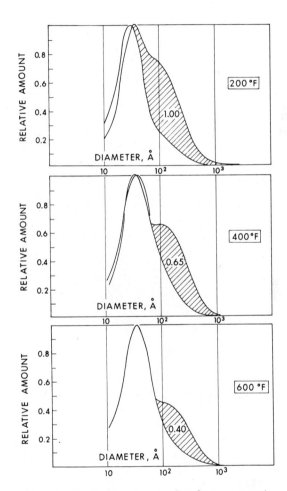

Figure 6.   GPC molecular size distributions at various temperatures

TABLE VIII

ASPHALTENE EXCLUSION EXPERIMENTS
AT 212°F AND 1 ATMOS.

| | CATALYST | | | |
	A	R	T	D
CATALYST WEIGHT, g	24.00	23.00	25.50	26.60
RESID WEIGHT, g	36.49	35.00	35.07	47.00
ASPHALTENE CONTENT				
INITIAL, %	6.3	6.3	6.3	6.3
FINAL, OBS., %	13.8	12.8	12.0	13.4
PORE VOLUME, ml/g	0.50	0.52	0.47	1.15
MICROPORE VOLUME, ml/g	0.42	0.50	0.44	0.45
MODE DIAMETER, Å	65	60	75	95

TABLE IX

METALS DISTRIBUTION IN RESID
FRACTIONS

	Ni (PPM)	V (PPM)
FEED	9	40
197°F INTERNAL	4	18
197°F EXTERNAL	15	58
403°F INTERNAL	2	48
403°F EXTERNAL	12	35
603°F INTERNAL	5	49
603°F EXTERNAL	10	27

mode diameters of 60-95 Å.

6.   Exclusion of asphaltenes and/or large molecules is also observed at 200, 400 and 600°F by similar technique involving GPC. The effect diminishes moderately with increasing temperature due to thermal dissociation into smaller particles.

7.   The exclusion of asphaltenes is matched by the distribution of nickel inside and outside the catalyst pore structure. Vanadium distribution is inconsistent, probably influenced by coke deposition at the higher temperatures.

ACKNOWLEDGEMENT

The authors acknowledge Dr. Dennis L. Saunders for the GPC analytical work and many useful discusssions.

LITERATURE CITED

(1)   Larson, O. A., Pittsburgh Catalysis Society, Spring Symposium, (April, 1972).
(2)   US 3,509,044 (Esso)
(3)   US 3,531,398 (Esso)
(4)   US 3,563,886 (Gulf)
(5)   UK 1,122,522 (Gulf)
(6)   NPA 6,815,284 (Hydrocarbon Research)
(7)   German 1,770,996 (Nippon Oil)
(8)   Drushel, H. V., Preprints, Div. Petrol. Chem., ACS, 17, No. 4, F-92 (1972).
(9)   Dickie, J. P., Yen, T. F., Preprints, Div. Petrol. Chem., ACS, 12, No. 2, B-117 (1967).

10

The Structural Form of Cobalt and Nickel Promoters in Oxidic HDS Catalysts

R. MONÉ and L. MOSCOU

Akzo Chemie Nederland bv, Research Centre Amsterdam, P. O. Box 15, Amsterdam, The Netherlands

Hydrodesulfurization catalysts consist of CoO and MoO_3 or NiO and MoO_3 on a γ-Al_2O_3 carrier (1-3). The catalysts are produced in the oxidic form; their actual active state is obtained by sulfiding before or during usage in the reactor.

Molybdenum oxide - alumina systems have been studied in detail (4-8). Several authors have pointed out that a molybdate surface layer is formed, due to an interaction between molybdenum oxide and the alumina support (9-11). Richardson (12) studied the structural form of cobalt in several oxidic cobalt-molybdenum-alumina catalysts. The presence of an active cobalt-molybdate complex was concluded from magnetic susceptibility measurements. Moreover cobalt aluminate and cobalt oxide were found. Only the active cobalt molybdate complex would contribute to the activity and be characterized by octahedrally coordinated cobalt. Lipsch and Schuit (10) studied a commercial oxidic hydrodesulfurization catalyst, containing 12 wt% MoO_3 and 4 wt% CoO. They concluded that a cobalt aluminate phase was present and could not find indications for an active cobalt molybdate complex. Recent magnetic susceptibility studies of the same type of catalyst (13) confirmed the conclusion of Lipsch and Schuit.

Schuit and Gates (3) proposed a model for the oxidic catalyst, in which the promotor ions lay just below the molybdate layer. These cobalt ions would stabilize the molybdate layer on the catalyst surface.

The promoting action of cobalt on the activity for hydrodesulfurization has been shown already in the pioneering work of Byrns, Bradley and Lee (14). This promoting action might be linked with the sulfiding step, since the actual catalyst is the sulfided form of cobalt- or nickel-molybdenum-alumina. Voorhoeve and Stuiver (15) and Farragher and Cossee (16) demonstrated the promoting action for the unsupported Ni-WS_2 system. Their intercalation model was based on these experiments.

It is expected that the activator system in the oxidic catalyst has a structure from which the right configuration is obtained on sulfiding. We feel that the promotor ions have to be

present in the neighbourhood of the surface molybdate layer and
that the assumption of a cobalt-aluminate phase might be an insuf-
ficient description of the oxidic hydrodesulfurization catalyst.
Therefore we have examined the presence of interactions between
cobalt and molybdenum from which a more detailed picture of the
oxidic catalyst can be deduced. We have studied therefore UV-VIS
reflectance spectra of adsorbed pyridine, a technique that has
been used before for this type of catalysts by Kiviat and Petrakis
(19).

Experimental Methods.

Catalyst Preparation. The catalysts were prepared by impregnation of γ-alumina extrudates (SA=253 m^2/g). Each impregnation
was followed by drying overnight at 120 °C and calcination at the
indicated temperatures during one hour. Molybdenum was brought on
the support as an ammonium molybdate solution; cobalt and nickel
as nitrate solutions. Each component was impregnated separately.
The impregnation of cobalt on the blanc support took place stepwise (1 wt% CoO in each step).

Catalysts. MoCo-124: Alumina was impregnated first with
molybdenum, then calcined at 650 °C, followed by the cobalt impregnation. The final calcination was varied from 400 - 700 °C. Composition: 12 wt% MoO_3 and 4 wt% CoO.
MoCo-122 and MoCo-123 were obtained by impregnation as carried
out for MoCo-124. Only 2 and 3 wt% CoO resp. were brought on the
catalyst. Final calcination temperature 650 °C.
CoMo-124: Alumina was impregnated first with cobalt stepwise.
The sample was dried at 120 °C and calcined at 650 °C after each
impregnation step. Afterwards the catalyst was impregnated with
molybdenum. Final calcination temperature 650 °C. The composition
was the same as for MoCo-124.
CoMo-124 B: Cobalt was brought on uncalcined boehmite, 4 wt%
in one step. Afterwards the sample was calcined at 650 °C, impregnated with molybdenum oxide (12 wt%) and calcined at 650 °C for a
second time. The surface area was 241 m^2/g.
MoCo-153: Alumina was impregnated first with molybdenum, dried
and calcined at 650 °C. Then the cobalt was brought hereupon. Two
final calcination temperatures were applied, 480 and 650 °C. The
composition was 15 wt% MoO_3 and 3 wt% CoO.
NiMo-124: The catalysts were prepared according to the Dutch
Patent 123195 (17). The alumina carrier was impregnated first with
nickel. Two final calcination temperatures were applied, 480 and
650 °C. These samples were impregnated with molybdenum. Final calcination temperatures 480 and 650 °C. The composition of the four
catalysts, which were obtained was 12 wt% MoO_3 and 4 wt% NiO, The
catalysts are indicated by the applied calcination temperatures
after each impregnation; e.g. NiMo-124 480/480.
MoNi-153: Molybdenum was brought on the alumina carrier first,

(calcination temperature 650°C), followed by the nickel impregnation. The final calcination temperature was varied. The composition was 15 wt% MoO_3 and 3 wt% NiO.

Reflection Spectroscopy. The reflection spectra were recorded with an Optica Milano CF 4 spectrophotometer, using magnesium oxide as the reference. The samples were milled in a mortar during 20 minutes. The reflectance did not change significantly when this milling time was increased. The spectra of the cobalt containing samples are shown, as they have been recorded (% reflectance against wavelength). The spectra of the nickel containing samples are plotted as a remission function (18) against wavenumber.

Infrared Spectroscopy. The procedure of Kiviat and Petrakis (19) was followed for the greater part. The spectra of adsorbed pyridine were recorded with a Perkin Elmer 621 spectrophotometer. The discs (15 mg/cm^2) were pressed in a RIIC die, using a pressure of 1000 kg/cm^2. These discs were rehydroxylated first on standing in humid air during two days before they were placed in the IR cell. Outgassing took place at 420°C. Pyridine was adsorbed at room temperature and the excess was removed by evacuation at 150, 250 and 350°C during one hour.

Pore structure. The surface area and pore volume were determined by N_2 adsorption. No significant changes were observed when the catalyst was calcined in the temperature region of 400 - 700°C indicating that no structural collapse took place in this temperature range.

Results.

Visible Reflection Spectra. The final calcination temperature of MoCo-124 samples has been varied in order to study its influence on the coordination of the cobalt ions. The reflection spectra are shown in Figure 1. The spectra of MoCo-124, calcined at 400 and 500°C show a broad absorption band, covering the whole spectral region, with a weak superposition of the characteristic triplet of cobalt aluminate. This indicates that the cobalt ions are for the greater part still on the catalyst surface and not in the alumina lattice. The spectra of the MoCo-124 samples, calcined at 650-700°C show a strong increase in intensity of the triplet band, while the broad absorption band has disappeared. This indicates the formation of a cobalt aluminate phase.

Infrared Spectra. Figure 2 shows the spectra of pyridine adsorbed on γ-alumina. Two types of Lewis acid sites are present; strong Lewis acid sites, which still bind pyridine on evacuation at 350°C and characterized by the 1622 and 1454 cm^{-1} bands and weak Lewis acid sites, characterized by the 1614 and 1450 cm^{-1} bands. Brönsted acid sites, which have characteristic bands around

1636 and 1540 cm^{-1} (19,21) are not observed in this spectrum. The
spectrum of pyridine adsorbed upon cobalt-alumina (4 wt% CoO) is
shown in Figure 2d. No change in the pyridine spectrum is observed
in comparison with the spectrum of pyridine on γ-alumina, indicating that the surface acidity is not markedly changed. A same
behaviour has been observed for nickel alumina. These results
agree with those of Kiviat and Petrakis (19).

The adsorption of pyridine on molybdenum-alumina (12 wt%
MoO$_3$) has been investigated for samples in both original and
rehydroxylated form. The spectra are shown in Figure 3. It appears
that Brönsted acid sites, characterized by the 1636 and 1540 cm^{-1}
bands are observed only, when the discs are rehydroxylated in wet
air before the calcination under high vacuum in the IR cell takes
place (Figure 3b). By consequence all spectra have been recorded
for such rehydroxylated samples. Only one Lewis band is observed
for the molybdenum-alumina sample, opposite to the observations
of Kiviat and Petrakis (19), who have observed two Lewis bands for
their samples.

Spectra of adsorbed pyridine have been recorded for the
MoCo-124 catalysts, for which the final calcination temperature
after the cobalt impregnation has been varied. It turns out that
the 400 and 500°C calcined samples and the 650 and 700°C calcined
samples show very similar spectra. Therefore we show only the
spectra of the 400°C (low calcined) and the 650°C (high calcined)
samples. Figure 4 shows spectra after desorption at 150 and 250°C.
Few Brönsted acid sites are observed in the low calcined MoCo-124
samples. The reflection spectra (Figure 1) indicate for these low
calcined samples the presence of cobalt on the catalyst surface,
because no cobalt aluminate phase could be detected. The high
calcined samples do show the presence of Brönsted acid sites; the
presence of a cobalt aluminate phase is concluded from the reflection spectra (Figure 1) for these samples.

These experiments indicate that at low calcination temperatures the cobalt ions are present on the catalyst surface and
neutralize the Brönsted acid sites of the molybdate surface layer.
At the higher calcination temperatures, the cobalt ions move into
the alumina lattice. The Brönsted acid sites reappear, indicating
that the situation on the molybdate surface is restored.

However, the molybdenum-alumina and the high calcined cobalt-
molybdenum-alumina samples still show an important difference. The
pyridine spectra of MoCo-124 indicate a second Lewis acid site,
characterized by the 1612 cm^{-1} band. This band differs from the
weak Lewis acid sites of the alumina support (1614 cm^{-1}) because
the position is significantly different. It also appears that the
strength of the bond between pyridine and the catalyst is stronger,
for the 1612 cm^{-1} band is still present after evacuation at 250°C,
while the weak Lewis band (1614 cm^{-1}) of the alumina has disappeared at this desorption temperature. Obviously the second Lewis
band for the MoCo-124 catalyst is introduced by the interaction of
cobalt with the surface molybdate layer. This interaction is

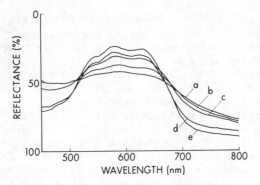

Figure 1. Reflectance spectra of MoCo-124 catalysts, calcined at different temperatures. a) 400°C, b) 500°C, c) 600°C, d) 650°C, and e) 750°C.

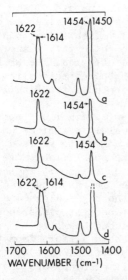

Figure 2. Spectra of adsorbed pyridine a) on γ-Al₂O₃ evacuated 1 hr, 150°C; b) as a), evacuated 1 hr, 250°C; c) as a), evacuated 1 hr, 350°C; d) on CoO-γ-Al₂O₃, evacuated 1 hr, 150°C

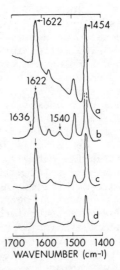

Figure 3. Spectra of adsorbed pyridine a) on MoO₃-γ-Al₂O₃, evacuated 1 hr, 150°C; b) on rehydroxylated MoO₃-γ-Al₂O₃, evacuated 1 hr, 150°F; c) as b), evacuated 1 hr, 250°C; d) as b), evacuated 1 hr, 350°C

clearly still present in the high calcined catalysts, where cobalt is present in the alumina lattice.

Reversed Impregnation. The reversed impregnation has been studied too. Impregnation of 4 wt% CoO on γ-alumina in one step generally leads to the formation of Co_3O_4 on calcination (black extrudates (22,23)). A stepwise impregnation results in cobalt aluminate formation (blue extrudates), showing its characteristic triplet in the reflection spectrum (Figure 5a). Impregnation of molybdenum on the cobalt containing support does not influence the reflection spectrum of the cobalt ions, as shown in Figure 5b.

Spectra of pyridine, adsorbed on this CoMo-124 sample are shown in Figure 6. The second Lewis band (1612 cm^{-1}) is present, indicating that the interaction between the cobalt ions and the surface molybdate layer is present too.

Boehmite Based Catalysts. Hedvall (24) has discussed the formation of cobalt aluminate from CoO and Al_2O_3. He has shown that a relatively fast solid state reaction takes place when the alumina undergoes a phase change, viz. $\gamma-Al_2O_3 \rightarrow \alpha-Al_2O_3$. This phenomenon is known as the Hedvall effect.

Such an effect might be expected when boehmite supported cobalt is being calcined, viz. during the phase transition AlO(OH) $\rightarrow \gamma-Al_2O_3$. Figure 7 shows spectra of pyridine, adsorbed on the sample CoMo-124 B, which has been prepared in this way. Spectra for MoCo-122, -123 and -124, containing 2, 3 and 4 wt% CoO resp. are shown for comparison. All these catalysts have had a final calcination of 650°C. Comparison of the spectra of CoMo-124 B and MoCo-124 indicates that the intensity of the 1612 cm^{-1} band, which is introduced by the interaction of the cobalt ions and the molybdate layer, is lower for CoMo-124 B than for MoCo-124. The spectrum for CoMo-124 B resembles that of CoMo-123, indicating that a part of the cobalt ions does not participate in this interaction.

Nickel Promoted Catalysts. Nickel containing catalysts are known to be sensitive for too high temperatures. The Dutch patent 123195 (17) claims that active nickel-molybdenum-alumina catalysts are obtained, when nickel is impregnated first. The calcination is critical however. According to this patent, catalysts calcined at 480°C are twice as active as catalysts, calcined at 650°C. Catalysts NiMo-124 have been prepared according to this patent and have been investigated. The hydrodesulfurization activity showed indeed a pronounced decrease on calcination at 650°C in comparison with the 480°C calcined sample.

The spectra of adsorbed pyridine are shown in Figure 8. The spectrum of the sample NiMo-124 480/480, which has been calcined twice at the lowest calcination temperature shows a clear 1612 cm^{-1} Lewis band. This band has a far weaker intensity for the 650/650 catalyst. The 480/650 sample shows a more intense 1612 cm^{-1} band than the 650/650 and 650/480 samples. This might indicate that the

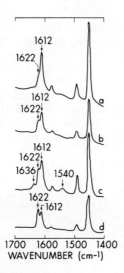

Figure 4.   Spectra of adsorbed pyridine a) on MoCo-124, final calcination 400°C, evacuated 1 hr, 150°C; b) as a), evacuated 1 hr, 250°C; c) on MoCo-124, final calcination 650°C, evacuated 1 hr, 150°C; d) as c), evacuated 1 hr, 250°C

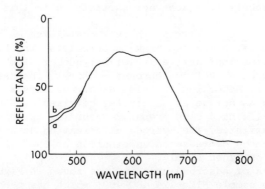

Figure 5.   Reflectance spectra: a) CoO-γ-Al₂O₃, calcined at 650°C; b) of CoMo-124, final calcination, 650°C

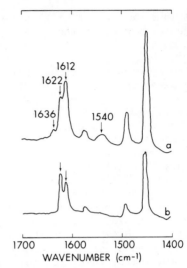

Figure 6. Spectra of adsorbed pyridine a) on CoMo-124, evacuated 1 hr, 150°C; b) as a) evacuated 1 hr, 250°C

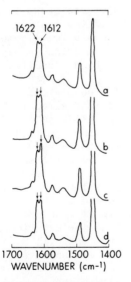

Figure 7. Spectra of adsorbed pyridine a) MoCo-122; b) MoCo-123; c) MoCo-124; d) CoMo-124 B. All after evacuation 1 hr, 150°C.

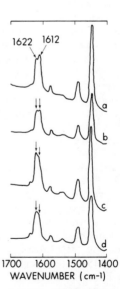

Figure 8. Spectra of adsorbed pyridine NiMo-124: a) 480/480; b) 480/650; c) 650/480; d) 650/650. All after evacuation 1 hr, 150°C.

nickel ions, which are still present in the neighbourhood of the
alumina surface after the first calcination at 480 °C, remain tied
to the molybdate layer to some extent after the second calcination
at 650 °C.

The spectra of adsorbed pyridine for MoCo-153 and MoNi-153
are compared in Figure 9. Two final calcination temperatures have
been applied, 480 and 650 °C. The spectra of the 480 °C samples (Figure 9a and 9b) are nearly identical. The Brönsted acid bands are
weak, while the 1612 cm^{-1} Lewis bands are strong. The intensity
of the Brönsted acid bands increases for both 650 °C calcined samples (Figure 9c and 9d). The Lewis acid bands show a marked difference now. The 1612 band remains high in intensity for the MoCo-
153 catalyst, but this Lewis band decreases appreciably in intensity for the MoNi-153 catalyst.

Reflection spectra for the MoNi-153 catalysts are shown in
Figure 10. The 480 °C calcined catalyst shows the characteristic
absorption band (25) of octahedrally coordinated nickel ions. The
650 °C calcined catalyst shows the characteristic spectrum of
nickel aluminate. These reflection spectra indicate that the
nickel ions migrate from the catalyst surface into the alumina, as
has been observed also for the cobalt-molybdenum-alumina catalysts.

Discussion.

Russell and Stokes (9) and Sonnemans and Mars (11) have presented strong evidence for the formation of a molybdate monolayer. It
appears from their experiments that each molybdate group covers
20-25 Å^2 of the alumina surface.The surface of the alumina support,
which has been used in this study, is high enough for a complete
spreading of 15 wt% MoO$_3$, so no bulk molybdenum oxide is expected
to be present. This has been confirmed by X-ray diffraction
measurements.

The molybdate surface layer in the molybdenum-alumina samples
is characterized by the presence of Brönsted acid sites (1545
cm^{-1}) and one type of strong Lewis acid sites (1622 cm^{-1}). Cobalt
or nickel ions are brought on this surface on impregnation of the
promotor. The absence of Brönsted acid sites is observed for both
cobalt and nickel impregnated catalysts, calcined at the lower temperatures (400-500 °C). Also a second Lewis band is observed at 1612
cm^{-1}.The reflection spectra of these catalysts indicate that no
cobalt or nickel aluminate phase has been formed at these temperatures. This indicates that the cobalt and nickel ions are still
present on the catalyst surface and neutralize the Brönsted acid
sites of the molybdate layer. These configurations will be called
"cobalt molybdate" and "nickel molybdate" and are shown schematically in Figure 11a.

For the high temperature calcined cobalt-molybdenum-alumina
catalysts, the presence of a cobalt aluminate phase has been concluded from the reflection spectra. The Brönsted acid sites reappear in the spectrum of absorbed pyridine, indicating that the

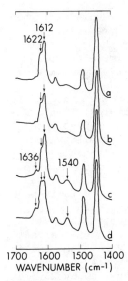

Figure 9. *Spectra of adsorbed pyridine on Mo-Co-153, final calcination: a) 480°C; c) 650°C; Mo-Ni-153, final calcination: b) 480°C; d) 650°C*

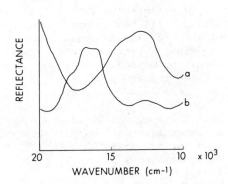

Figure 10. *Reflectance spectra of MoNi-153: a) final calcination, 480°C; b) final calcination, 650°C*

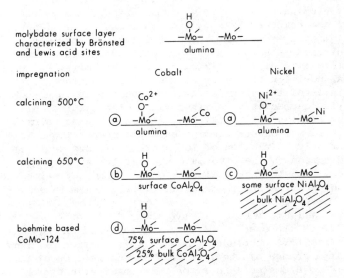

Figure 11. *Schematic representation of the different surface structures*

bond between cobalt and the molybdate layer is broken. Still the presence of two types of Lewis acid sites is observed (1622 and 1612 cm^{-1}). This suggests that no bulk aluminate phase is formed, but that the cobalt ions remain present in the first layer of the alumina lattice, still having an interaction with the surface molybdate layer.

The optimal activity for a cobalt-molybdenum-alumina catalyst is obtained by calcination at the higher temperatures. This means that the cobalt ions, present as a cobalt aluminate phase according to the reflectance spectra and the magnetic susceptibility measurements, still have a pronounced promoting action after this calcination. The assumption of cobalt present in the surface layer of the alumina lattice explains both the high activity due to the cobalt promotion as well as the presence of the second Lewis band. This configuration is shown schematically in Figure 11b.

Some bulk cobalt aluminate formation is expected to take place for the boehmite based catalyst, owing to a Hedvall effect (24).The spectrum of adsorbed pyridine on CoMo-124 B shows indeed a weaker 1612 cm^{-1} band, comparable with the intensity of this band for the MoCo-123 catalyst. This indicates that about 25 % of the cobalt ions has disappeared in the bulk of the alumina. (Figure 11d).

The reappearance of Brönsted acid sites has been observed for the high calcined nickel-molybdenum-alumina catalysts. The presence of a nickel aluminate phase has been concluded from the reflectance spectra. The second Lewis band (1612 cm^{-1}) has a very low intensity, in comparison with the cobalt containing catalysts of a same composition and after the same calcination conditions.

Nickel aluminate is a partly inverse spinel (26) in which the nickel ions occupy both tetrahedral and octahedral holes, opposite to cobalt aluminate, which is a normal spinel. Romeyn (26)has shown for the compound $NiAl_2O_4$ and Lo Jacono et.al. (25) for nickel alumina systems that the nickel ions occupy only a fraction of the tetrahedral sites, never exceeding 25 %. The strong reduction in intensity of the 1612 cm^{-1} Lewis band might be due to the octahedral position of the greater part of the nickel ions in the surface layer of the alumina lattice. However it is also known that the activity decreases strongly for the high calcined nickel-molybdenum-alumina catalysts (17), at temperatures normally applied in the cobalt-molybdenum-alumina catalyst preparation. This indicates that only a fraction of the nickel ions are available for a promoting action in the high calcined catalysts. We suggest that this is due to bulk nickel aluminate formation, by which the nickel ions are lost in the alumina lattice. This is schematically shown in Figure 11c.

Kiviat and Petrakis (19) have shown that the cobalt and nickel ions influence the spectra of pyridine adsorbed on molybdenum-alumina. They have concluded that the introduction of these ions results in a change of the ratio of the intensities of the bands of the Lewis and Brönsted acid sites. Our observations show that this ratio is strongly dependent on the calcination temperature. This temperature determines the interaction between the surface molybdate layer and the promotor ions and might be correlated with the hydrodesulfurization activity of the catalysts.

Concerning the activity of the cobalt-molybdenum-alumina catalysts, it is the experience in this laboratory that the optimal activity is obtained by calcination at high temperatures. The high calcined form is probably preferred, because in this form the cobalt ions are hidden in the first layer of the alumina lattice and are not very susceptible for reducing conditions in this configuration. It is well known that this is the case for the (low calcined) nickel-molybdenum-alumina catalysts. This might be expected from the far more exposed form of the nickel ions (Figure 11a). No correlations between activity and the presence of a cobalt aluminate complex as indicated by Richardson (12) can be concluded. The low calcined cobalt-molybdenum-alumina catalysts do show the presence of a "cobalt molybdate" phase; however this phase disappears on calcination at 650-700°C, where the catalysts still have a high activity. This shows that the cobalt ions are not lost in the alumina lattice.

Conclusions.

It has been found by measuring spectra of adsorbed pyridine on various $CoO-MoO_3-Al_2O_3$ systems that there is a clear interaction of the cobalt ions with the $MoO_3-Al_2O_3$ surface layer. This conclusion is based upon the occurence of a second Lewis band in the spectrum of adsorbed pyridine, which has been observed in the whole region of the calcination temperatures applied.

A picture has been formed of the way in which the promotor ions are built in the $MoO_3-Al_2O_3$ system. The neutralization of the Brönsted acid sites, as originally present in $MoO_3-Al_2O_3$ systems by the cobalt ions for the catalysts calcined at low temperatures (~ 500°C) indicates that the cobalt ions are present on the catalyst surface. The liberation of these sites in catalysts calcined at high temperatures (~ 650°C) and the observation of the characteristic reflectance spectrum of $CoAl_2O_4$ show that the cobalt ions enter the alumina lattice. However the interaction between cobalt and molybdenum, as indicated by the second Lewis band remains present. This leads to the conclusion that the cobalt ions are present in the surface layers of the alumina lattice.

Similar interactions have been observed for the nickel promoted catalyst. However, the degree of interaction depends on the calcination temperature. This interaction disappears for a great part at increasing temperatures. This is ascribed to bulk nickel aluminate formation.

These observations might contribute to the understanding of the differences in catalytic performance between cobalt and nickel promoted hydrodesulfurization catalysts.

Literature Cited.

1) McKinley, J.B., in "Catalysis", P.H. Emmett,(ed), Vol 5,
   p 405, Reinhold, New York 1957.
2) Schuman, S.C. and Shalit, H., Catal. Rev., (1970),4, 245.
3) Schuit, G.C.A. and Gates, B.C., AIChE, (1973),19, 417.
4) Seshadri, K.S. and Petrakis, L., J.Catal. (1973),30, 195.
5) Massoth, F.E., J. Catal. (1973),30, 204.
6) Dufaux, M., Che, M. and Naccache, C., J. Chim. Phys. Physico-
   chim. Biol., (1970),67, 527.
7) Masson, J. and Nechtstein, J., Bull. Soc. Chim. Fr., (1968),
   3933.
8) Miller, A.W., Atkinson, W., Barber, M. and Swift, P., J. Catal.
   (1970),22, 140.
9) Russell, A.S. and Stokes, J.J., Ind. Eng. Chem., (1946),38,
   1071.
10) Lipsch, J.M.J.G. and Schuit, G.C.A., J. Catal., (1969),15, 174.
11) Sonnemans, J. and Mars,P., J. Catal., (1973),31, 209.
12) Richardson, J.T., Ind. Eng. Chem. Fundamentals, (1964),3, 154.
13) Roelofsen, J.W., unpublished results.
14) Byrns, A.C., Bradley, W.E., and Lee, M.W., Ind. Eng. Chem.,
    (1943), 35, 1160.
15) Voorhoeve, R.J.H. and Stuiver, J.C.M., J. Catal., (1971),23,
    228,236,243 (1971).
16) Farragher, A.L. and Cossee, P.,Proc Fifth Intern. Congr.
    Catal., Palm Beach, p 1301, North-Holland Publishing Company,
    1973.
17) Chevron Researh Company, Dutch Patent 123195.
18) Wendlandt, W.Wm. and Hecht, H.G., "Reflectance Spectroscopy",
    Interscience Publishers, New York, 1966.
19) Kiviat, F.E. and Petrakis, L., J. Phys. Chem.,(1973),77, 1232.
20) Parry, E.P., J. Catal., (1963),2, 317.
21) Grimblot, J., Pommery, J. and Beaufils, J.P., C.R. Acad. Sci.,
    Ser. C., (1973),276, 331.
22) Tomlinson, J.R., Keeling, R.O., Rymer, G.T. and Bridged, J.M.,
    Actes du deuxième Congr. Intern. de Catalyse, p 1831, Ed.
    Technip, Paris (1961).
23) Nishina, R., Yonemura, M. and Kotera, Y., J. Inorg.Nucl.Chem.,
    (1972),34, 3279.
24) Hedvall, J.A. and Leffler, L.,Z. Anorg. Allg. Chem., (1937),
    234, 235.
25) Lo Jacono, M., Schiavelo, M. and Cimino, A., J. Phys. Chem.,
    (1971),75, 1044.
26) Romeyn, F.C., Thesis, University of Leiden, 1953.

INDEX

INDEX